MANUAL OF REMOTE WORKING

MANUAL OF REMOTE WORKING

Kevin Curran and Geoff Williams

Gower

© TNL 1997

All rights reserved. No part of this publication may be reproduced, stored in a retrieval system, or transmitted in any form or by any means, electronic, mechanical, photocopying, recording or otherwise without the permission of the publisher.

Published by
Gower Publishing Limited
Gower House
Croft Road
Aldershot
Hampshire GU11 3HR
England

Gower
Old Post Road
Brookfield
Vermont 05036
USA

Kevin Curran and Geoff Williams have asserted their right under the Copyright, Designs and Patents Act 1988 to be identified as the authors of this work.

British Library Cataloguing in Publication Data
Curran, Kevin
Manual of remote working
1. Telecommuting - Handbooks, manuals, etc.
I. Title II. Williams, Geoff
331.2'5

ISBN 0 566 07839 2

Library of Congress Cataloging-in-Publication Data
Curran, Kevin 1954–
Manual of remote working / Kevin Curran and Geoff Williams.
p. cm.
Includes bibliographical references and index.
ISBN 0–566–07839–2 (cloth)
1. Telecommuting. 2. Home laboar. I. Williams, Geoff, 1954–
II. Title
HD2336.3.C87 1997
658.3—dc20 96–32169
CIP

Typeset in Century Old Style by Raven Typesetters, Chester and printed in Great Britain by Biddles Ltd, Guildford.

Contents

Preface

This is not a textbook! It is not an academic treatise, nor just a simplistic series of 'How to . . . ' bullet-points.

Our aim in this Manual is to present the key issues facing those considering setting up or wishing to improve a *remote working* (sometimes known as *teleworking*) scheme and to provide checklists of best practice, and to support these with examples of actual documents and systems. The advice given here is based not only on considerable desk research on what is happening around the world in remote working, but also on personal experience of running a company (TNL) in which all the personnel are remote workers. In fact much of the supporting material shown here as examples is used in the day-to-day operation of TNL.

We have not set out to provide the answers to everything! There is no *one* way of running any business, let alone a remote working business. The reader will have to *interpret* what is suggested here. Some fundamental questions need to be answered – not by us, but by the potential managers of a remote working scheme. We hope we have at least posed the key questions.

We suggest you look first at Chapter 1 for an indication of what is involved. Where you turn to next is optional. While the content is presented in the order of events a person might go through to plan, set up and manage a remote working scheme, you must decide for yourself just where to dip in.

The Manual is designed to help different sorts of reader. It will be invaluable to the senior manager of a large company establishing a remote working scheme for a group of regular workers, or for the trainer or consultant who may be advising the senior manager. It should also prove an excellent resource for anyone planning to set up a brand-new company based on remote working.

Kevin Curran
Geoff Williams

Acknowledgements

Kevin Curran and Geoff Williams would like to acknowledge and thank the following people for their contributions to this Manual:

David and Krysia Thompson
Andrew Mitchell
Alison Ferry.

Without their help it would not have been possible.

Glossary of terms

For the purpose of this Manual we aim to be consistent about the terms we use. Readers may have a different understanding of them, so we list them here as we understand them.

Co-worker: we often talk about our working colleagues as 'internal customers' since they may be recipients of our work; in this Manual, we prefer 'co-worker' since remote workers are more likely to form collaborative, if temporary, partnerships; a person working on the same project.

Customer: the person who commissions work to the organisation; he/she may have a contract with the employer to undertake work, or the employer may be the customer; he/she is the end-user of the work done by the worker.

Employer: the person who pays for the work to be done by the worker; the employer may be the legal employer, contractor or agent; the organisation which employs remote workers, either as part of its work activities or as an agency which solely employs remote workers.

Manager: the person, either an employer or someone appointed by the employer, to take the role of manager; in corporate schemes, this may be a regular managerial role; in an agency or partnership, it may be a project manager role or someone allocated certain managerial responsibilities in addition to their own remote working job; a supervisory role.

Remote working: a work practice where jobs are carried out in a place other than an office or traditional work setting, using as much suitable equipment as possible and practicable for a given job, with a significant role for information technology in communications; the synonyms or near-synonyms of remote working are: teleworking, homeworking, off-site working, telecommuting.

SMART (targets): Specific, Measurable, Achievable, Relevant, Time-based.

Telecommuting: usually understood to mean the sort of work done by a regular employee who works mostly at home to reduce commuting.

Worker: the remote worker, who may be a professional, semi-skilled or unskilled, and who may be a legal employee, a sub-contractor or a freelance worker.

1 Introduction to Remote Working

- Introduction
- What makes remote working special?
- Jobs suitable for remote working
- Benefits of remote working for the worker
- Benefits of remote working for the employer
- Disadvantages of remote working for the worker
- Disadvantages of remote working for the employer

INTRODUCTION

We have called the subject of this Manual **remote working** as an all-embracing term for teleworking, homeworking, telecommuting, and so on. Each of these has its own special features which are not discussed here. We want instead to focus on the central issues *common to them all* which, if properly approached, should ensure remote working is effective.

Remote working usually means:

> *Working from a place other than an office or other traditional work setting, using as much suitable equipment as is possible and practicable for a given job, with a significant role for information technology in communications.*

There are many examples of organised and *ad hoc* remote working, from large-scale corporate schemes aimed at reducing overheads and travel time, and to improve the quality of workers' lives, to the occasional taking off of a week to work quietly at home on that essential sales presentation. It is by no means new, nor is it uncommon.

The most common forms of *organised* remote working are:

- being a regular, salaried employee of an organisation, but often working away from the office. A typical example would be a reporter, or a secondee on a project with a subcontracting firm. Such workers tend to be professionals, perhaps with professional qualifications, and they often enjoy all the benefits of regular employees, whether permanent employees or on a fixed-term contract. They may be working away from their normal office only temporarily, and may at times work as much in the office as away from it.
- being a regular, piece-work or hourly paid employee, working at home. A typical example would be a data-entry clerk, or a telephone-based software support helper. Such workers may, or may not, be deemed professionals, but they are not always part of the permanent staff establishment of the

employer, even if they in fact only work for one employer. They may, or may not, have all the benefits of regular employees. They tend to be part of a pool of workers that allows the organisation to handle fluctuations in the quantity of work required.

- being a subcontracted, freelance worker, working for one or more employers on projects. The worker is self-employed and may even have formed him/herself into a limited liability company. The worker may deal direct with employers or through an agency or main contractor. An example of the former is the freelance total quality trainer running programmes for a company; an example of the latter is the systems analyst doing work for a software company who has won a contract to introduce a new application into another organisation. Such workers allow an organisation not so much to manage fluctuations in work quantity, but rather to bring in quality skills as and when needed.

It is clear from the above list that remote working is not new and that it is already quite common. It is also apparent that remote working cannot always be neatly slotted into one of the above categories. There are always overlaps. The specific form of remote working depends mainly on four key factors:

1. The required relationship between employer and worker.
2. The specific nature of the work to be done.
3. How the work is to be managed.
4. Who is to do the work.

There is perhaps an implied sequence in the above list that is worth examination. Most greenfield-site operations would probably define first the work that has to be achieved – i.e. the what, how, who – and then the formal policies that regulate the workers. A 'culture' is allowed to evolve (or is pointed towards) through the personal style of management and workers.

If moving a current section from its regular workbase to a remote working one, the sequence might be: who, then how, since the what, the nature of the work, has already been defined. Perhaps only then is consideration given as to the impact remote working may have on existing contractual relations.

Some organisations, especially telecottages and teleworking agencies, begin by defining first the relationship between employer and worker. As more and more fragmentation of the job market takes place and as telecomputing systems mature, and outsourcing increases, so will more workers come together – not physically, but in 'virtual companies' or temporary work partnerships – to work on projects. The relationship moves from employer/employee to contractor/contractee and, perhaps, partner/partner. This, in turn, impacts on how projects are defined and managed.

All four factors, listed above, have to be addressed, and it is probable they have to be addressed in light of one another.

Throughout this Manual, examples are given of working documents and systems: they are illustrative. It is important the reader 'translates' the examples given here to their context.

WHAT MAKES REMOTE WORKING SPECIAL?

If remote working is not new, what makes it worthy of special consideration? Surprisingly (or perhaps not so surprisingly), remote working leads to:

- more hours worked per day. There is no settling-down time at work after arrival, or around lunchtime or tea/coffee breaks, etc.; it is said to be normally on average 20–30 minutes per person per day.
- more work done per hour. There are no office distractions. Home still does offer distractions, but good remote workers are said to be more task-oriented people, so (hopefully) their work regime reduces home distractions. Selection procedures should take into account recruits' resistance and ways of coping with distractions at home.
- working at odd (to employers) hours. The organisation must accept people's preference for working certain hours which may be outside the traditional nine to five, Mondays to Fridays. Remote workers tend to optimise their work time, doing highly value-added work in quality time.
- less accidental absenteeism, less stress connected with commuting and occasional home circumstances holding back the worker (e.g. sick child, school days off, repairman awaited, etc.).

There are benefits (and consequences) involved in working remotely. In many ways, managing a remote working organisation is similar to managing any other organisation: it needs a business plan, objectives, operational and information systems, able workers, and effective management. The evidence suggests that all of these need to be strengthened, especially:

- the choice of remote worker
- the motivational problems associated with remote working
- the management of remote workers and the choice of manager

- communication practices, in particular computer-based communications.

These factors are considered in the following pages.

JOBS SUITABLE FOR REMOTE WORKING

The following are among the types of job that have been successfully transplanted into remote work:

1. *Professionals* – accountants; architects; in the field of management: marketing, PR, human resources and finance; and financial analysts and brokers, distance learning/ education and consultancy.
2. *Professional support* – bookkeeping, translating, research, proofreading, indexing, remote maintenance and data/ information retrieval.
3. *Field workers* – company reps, surveying, estate agency work, remote control, auditing, insurance brokerage and journalism.
4. *IT specialists* – systems analysts, software programming, software documentation, etc.
5. *Clerical support* – data entry, word processing, directory enquiry services and telesales.

It has been said that almost all jobs that do not require specialist/on-site equipment or laboratory or production-line teamwork can be altered to suit remote work. A job suitable for remote work should involve:

- a high degree of cerebral, rather than manual, work
- work done as an individual, or with clearly defined areas of individual work
- minimum direct supervision
- measurable deliverables
- no costly or bulky pieces of equipment.

BENEFITS OF REMOTE WORKING: WORKER

The benefits of remote working for workers include:

- Improved quality of working life – no constraints of conventional office-based circumstances.
- Reduction of stress associated with commuting, child/elderly care, home commitments and wasting time.
- Increased autonomy and personal flexibility which results in better planning and executing of work schedules (time and geographical flexibility).
- Extended range of employment opportunities – without the need for relocation – to the disabled or those living at a distance from work sites, with poor public transport network, etc.
- Working in a physically comfortable and familiar setting.

BENEFITS OF REMOTE WORKING: EMPLOYER

The benefits of remote working for employers include:

- A way of introducing flexibility in organisations (so

desirable now). A flexible company may contract and expand depending on the market demands.

- Reduction of overheads – reduction of office space (rent, electricity, heating, etc.), furniture, and so on.
- Increased productivity (remote workers claim they feel more self-motivated and do not experience the usual on-site interruptions), increased concentration and job satisfaction.
- Retention of skilled workers, as well as recruitment of remote workers with the right skills and without the need for relocation. Bridging the career gap, which is especially important for women with small children who only give up their jobs temporarily.
- Remote working is seen as part of a broader equal opportunities programme.
- Opportunities to provide employment to disadvantaged rural or urban areas (a more socially and politically correct benefit).
- An opportunity to expand without drastic increases in overheads.
- An opportunity for flexible use of resources.
- Access to a pool of skilled and experienced workers.
- More disciplined use of time and therefore increased productivity.

DISADVANTAGES OF REMOTE WORKING: WORKER

Disadvantages of remote working for workers include:

- Being out of touch with other professionals.

- Too much peace and quiet!
- Strain on relationships at home.
- Logistical difficulties with children, carers, etc.
- Missing out on the social aspects of regular working.
- Underutilisation of skills.
- Lack of feedback on performance.
- Lack of someone to turn to for counselling, guidance or advice.
- Feeling of being exploited (poor pay, no fringe benefits).
- Possible gaps between work and little job security.
- Little sense of career advancement or promotion.

These disadvantages are real ones and need managing. The employer can do much to minimize them through defining the relationship between employer and worker, so that both parties are under no illusions.

It is also in *both* parties' best interests if the employer can provide as many of the conventions of a regular employer – employee relationship as reasonably possible. How much the employer is prepared to 'invest' in training, offering work to fill in gaps or socialising, etc. will depend on the perceived returns in the future and probably also on the warmth of the relationship.

DISADVANTAGES OF REMOTE WORKING: EMPLOYER

Disadvantages of remote working for employers include:

- difficulties of motivating staff at a distance
- extra efforts needed to manage remote workers, especially the need to plan well

- loss of direct supervisory control.

The list is short since few of the disadvantages cannot be overcome by effective management.

2 Planning a Remote Working Operation

- Business issues
- Business plan
- Contractual relations
- Relationships between employing organisation, workers and customers

Appendices:

A. SWOT analysis

B. Cashflow pro forma

C. Mission statement

D. Employment contract checklist

E. Payment issues

Recommended reading

BUSINESS ISSUES

Demands on the remote working organisation include the need for all the usual underpinnings of any good business:

- business objectives
- detailed business plan
- customers and plans to manage customer relationships
- sales promotion, marketing (advertising internally, if appropriate, and externally, locally and nationally)
- detailed market research
- well-defined work practices, often in project management terms
- agreed forms and frequency of communications
- quality control procedures
- CVs of actual workers and people on file ('potentials' and 'casuals')
- skills profiles
- job design, job description and person specification
- map of skills/resources readily available (this is linked to work flow and allows organisations to grow or shrink quickly without heavy investment in growth, i.e. temporary growth).

There are many guides on how to plan a business, available from bookshops, banks and Business Advice Centres. We do not aim to duplicate these here, but merely highlight the key features.

In planning a remote working operation, whether as a separate business or as an extension of normal corporate activities, it is important to determine objectives: what precisely do you hope to achieve? An example is given in Appendix A of the objectives and thinking behind the setting up of TNL.

Planning and administration of a remote working operation is time-consuming, and employers need to recognise this and organise themselves to allow for the increase in management time (especially time spent on administrative matters) and intensity of responsibility, as in the following:

- linking the remote worker with customers: this may not always be necessary, especially if the remote worker is dealing directly with the customer, though even here there may be need for a brokerage role
- project management activities
- cash flow, cash projections, debts and procedures for dealing with them
- both technical and motivational support of workers.

External consultants/help may be called in to carry out the initial stages of these activities, as well as to provide training for management and workers to carry on remote working for themselves.

BUSINESS PLAN

Remote working, either as a separate business venture or a business within a larger organisation, needs its own detailed **business plan**.

It serves as a basis for:

- the work plan
- work scheduling
- seeking financial support
- budgeting

- human resource planning
- corporate image.

The **marketing plan** is of equal importance and is explained here in some detail, together with the business plan. What follows is a brief outline of what the main sections of a business plan should contain.

1. *Summary* – sums up the points that follow, i.e. corporate image, products/services to be offered, timescales, targeted market, competition, financial aspects and personnel requirements.
2. *Contents* – only for larger business plans (i.e. with page numbers) to locate topics quickly. Omit if less than 10 pages long.
3. *Introduction* – a clear statement of the business and its objectives; it should be short – just a few paragraphs. If the aim of the business plan is to raise finance, this should be clearly stated at this stage.
4. *Business history* – applies only if you are acquiring an existing business or planning a considerable expansion or diversification of your business; it should include the date the business started, together with dates of any significant recent events that influenced the business. This part should also include: the main product/service and the market, and what has made the business successful, as well as the present financial position.
5. *Personnel (management)* – details of proprietors/directors of the business, together with their age, qualifications, personal means (e.g. property ownership) and relevant experience (could be CVs). For first-time business owners it is useful to include business/management courses and any

advice or support from an enterprise agency or similar body. A statement of staff needs (at all levels) should be included – i.e. numbers, skills, likely wages, full-time/part-time and possible training requirements.

6. *Product/service* – a description of the product/service the business is offering, or its present stage of development if you are just starting. For service businesses a certain amount of detail has to be included: what exactly the service is, its market demand and the likely customers.
7. *Market research and marketing plan* – market research is vital; a business may have a very good product or service, but there may not be enough demand for it. Market research provides the basis for deciding if the product/service is likely to succeed. Findings of the market research should include:

   - the size of the market
   - presence (and advantages) of main competitors (provide names and details of the main competitors, together with any relevant intelligence)
   - what advantages your business has over its main competitors.

   Marketing is a continuous activity; it consists of:

   - gathering information
   - finding out what customers want
   - looking at new opportunities to sell the product/service.

   To carry out market research, you should *know your customers* – their requirements can change rapidly, so you need to monitor these closely:

- stay close to the customers to foster good relationships
- gather information on your product/service (How do customers see your company?)
- activities of your competitors and how they are seen
- look out for changes in the market
- talk to customers and *listen* to them
- read the press and relevant journals
- do some library research, if necessary.

You should also *know yourself*:

- try SWOT analysis (for a pro forma SWOT analysis sheet see Appendix A to this chapter; ask what are your company's strengths and weaknesses, what opportunities exist for you to take advantage of and what threats are there?
- are there niches in the market that you may be missing?
- are you missing out on customers?

From this research you can now devise your marketing strategy, which should consider:

- *product/service* – what are its main features, strong points, quality, image?
- *price* – is your pricing right, do customers see value in your product/service, any discounts/allowances, methods of payment?
- *promotion* – advertising, publicity, type of advertising.

All statements should be supported by figures, wherever possible.

8. *Suppliers/subcontractors* – list the ones you intend to use; it is important *not* to rely on just one source of services as problems of supply may result in a disaster for your

business. Always have a list of alternative suppliers, and list them all.

9. *Equipment* – state what equipment is needed: should it be purchased, leased/ hired or new or second-hand? Quote sources. It is especially important if the equipment is expensive.
10. *Financial aspects*

- *Pricing* – this specifies how you will price your product/ service. For services you must state hourly rates.
- *Cashflow forecasts* – these should be planned on a monthly basis, normally for the 12 months ahead. They show the estimates and assumptions, as well as facts and figures (for an example of a cashflow pro forma see Appendix B to this chapter).
- *Grants and other financial assistance already received* – these should be noted here; remember not to underestimate your overheads – it is better to be pessimistic!
- *Projected profit and loss* – usually prepared by an accountant. It should indicate the level of overheads and profits the business will make.
- *Projected balance sheets* – for projects requiring a start-up capital of £20 000 plus.
- *Risk assessment* – potential risk may come from new competitors, changes in legislation, sickness or fire, flood etc.; or simply the turnover level is lower than expected. It is another cashflow forecast minus some 20–50 per cent (with the overheads remaining the same). List any contingency plans for such eventualities.
- *Financial requirements* – statement of the total sum of money needed, together with how much is already

allocated (from other sources or from yourself) and what the needed money will be spent on.

11 *Implementation schedule* – a statement of timetable indicating dates for development and decisions and critical dates ('milestones').
12. *Corporate image* – a statement that aims to capture the 'style' of the business and how customers are intended to see the business, and how this image will be built and maintained.
13. *Mission statement* – has become popular over the past ten years; it attempts to capture the philosophy of the organisation, its purpose, objectives and values. Sometimes, because the aspirations in a mission statement are at odds with what happens in practice, they become derided. However, the management of a remote working operation should attempt to state what is important and use this to clarify expectations, both among remote workers and customers. For an example of a mission statement see Appendix C to this chapter, though you may want to include different things or sharpen your focus on certain aspects.

The business plan needs to be reviewed regularly, say, every six months. Business and market conditions can change rapidly and a common cause of failure of any business is that the management fail to monitor such change and fail to adjust their plans accordingly.

CONTRACTUAL RELATIONS

There are several different forms of relationship between employers and remote workers (these still need to be researched and some are still unclear or are affected by legislation):

1. Employer–employee (standard relationship).
2. Employer–employee (on a retainer).
3. Self-employed – the remote worker may rely on work from just one employer; there is a reasonable guarantee from the employer of the flow of work. The remote worker undertakes individual projects, and pay is per project (on the basis of the estimated number of hours) and negotiable. There may, or may not, be benefits and equipment provision. Or the remote worker works for more than one employer. The form of the relationship will be as above, though it is unlikely there will be benefits and equipment provision.
4 Strategic alliances – of partners or businesses.

Remote working companies may mix and match relationships according to circumstances. Companies need to be flexible in their use of different forms of these relationships – especially as the business grows and relations between worker and employer mature.

CONTRACTS WITH EMPLOYEES

If employing remote workers, as in a formal employer–employee relationship, contractual arrangements should include the following, discussed and agreed by both parties:

- place of work
- hours of work

- benefits, if any
- equal opportunity rights
- rate of pay
- job security (a retainer, guarantee of minimum work, etc.)
- working conditions
- work practices and rewards
- access to training/promotion
- career development
- health and safety
- discipline/grievance procedures, dismissal.

A checklist for a contract is provided in Appendix D to this chapter. For more information about the legal considerations see Chapter 8, p. 265.

One of the reasons for setting up a remote working operation is to enable an employer to have a pool of known workers readily available to pick up sudden demands in work. The flow of work may not always be steady. In such situations, the employer may consider paying a retainer to keep the workers on his/her books.

CONTRACTS WITH SELF-EMPLOYED REMOTE WORKERS

In situations where there is no formal employer–employee relationship, the remote worker is likely to be self-employed and paid on a contract basis.

Contracts can be hourly paid, day-rate or piece-rate – i.e. per unit of work or per contract. The types of contract may define:

- a constant flow of work
- work with deadlines and no continuity (heightened risks for remote workers with no steady income).

It *is* possible to have the same contractual arrangements as for conventional, on-site workers, it is at the discretion of employers. This can help reduce marginalisation.

It must be established within the contract what equipment and materials are needed to carry out the work. The basic equipment most likely to be needed by the remote worker includes:

- computer
- modem
- fax, or fax-modem
- photocopier
- telephone line (can be separate from domestic)
- filing cabinet.

The contract should stipulate the precise nature of the equipment, including the computer specifications, which over time tends to be upgraded. It must also be stipulated who owns what. If the equipment is loaned or leased, the terms (including the duration of the loan) must be specified and legally binding; it must be stated who pays for loss or damage and/or who pays insurance.

RELATIONSHIPS BETWEEN EMPLOYING ORGANISATION, WORKERS AND CUSTOMERS

In addition to the relation of management of the remote working operation and the remote workers themselves, the relationship with **customers** also needs to be defined.

Remote working may require the formal specification of the relationship between remote worker and customer, including:

- frequency and forms of contact with customers
- visits, locations of meetings, perhaps working from customers' premises
- areas and limits of direct negotiation between worker and customer on specifications and work progress
- rewards for bringing in new customers
- customer-oriented behaviour required of worker
- payment routes – customer to worker to employer, or customer to employer to worker (see Appendix E to this chapter for more about Payment Issues)
- payments for pre-contract work – e.g. initial meetings with the customer to discuss project
- role of worker in representing the corporate image to the customer.

The reasons for the above can be seen when one considers some of the potential pitfalls:

- Self-employed remote workers may work for more than one employer and may experience a conflict of interest.
- Remote workers, especially in service support roles, may come into frequent contact with customers, but without the close supervision of management.
- If self-employed, remote workers need to be clear how far they can act on management's behalf and what their decision-making powers are. As far as the customer is concerned, the remote worker *is* the company they are paying.
- Remote workers will in time develop their own customer contacts. If the relationship between remote worker and management is good, then this will be seen as of benefit to both sides. However, the remote worker may wish to set up business for him/herself. Or the remote worker may

expect added rewards for bringing in new customers to the business.

None of the above problems is unavoidable, if relationships are clarified up front.

Appendix A: SWOT analysis

BACKGROUND

Managers in organisations have, at any given time, an overall picture of what they see as the key issues facing their organisation. This will have involved them, formally or informally, in considering elements internal to the organisation (e.g. staffing, skills, finance, marketing, organisational structure, standards and service), as well as those factors impinging from outside (e.g. taxation, government policy and population movements).

Building up the picture requires thinking about a range of variables. For effective forward planning (including choice of future policies) such thinking is essential, but also vital is the need to place such thinking in a structured and usable form – i.e. through a 'guidance' frame. A SWOT analysis provides that frame. It places information gained from reviewing the internal and external environments in the form of: *S*trengths, *W*eaknesses, *O*pportunities, *T*hreats.

The note on p. 27 provides information as to what is involved in undertaking a SWOT analysis. Remember that individual perceptions and perspectives influence the SWOT, so that managers working in different parts of the same organisation are likely to have differing opinions, for example, as to whether a particular feature of the organisation represents a strength or weakness. For that reason, a cross-section of views can aid the final outcome.

WHAT DOES SWOT INVOLVE?

1. **Current policies**
   Identify the organisation's current:

   - values and beliefs
   - objectives (i.e. what it is trying to achieve)
   - strategies/policies (i.e. how it is seeking to achieve its objectives).

2. **Strengths and weaknesses**
   Look inside the organisation and ask:

   - What special attributes/distinctive abilities are possessed by the organisation?
   - What do customers most value?
   - What aspects of the organisation make it less effective?
   - What detracts from the product/service provided?

   The above questions involve looking at e.g.
   decision-making
   finance
   management style
   organisational structure
   standards
   etc.

   Resulting answers will identify the organisation's

   - Strengths, e.g. innovation
     financial situation
     customer base
     product/service
     etc.

- Weaknesses, e.g. unclear decision-making process
  marketing strategy
  debt
  etc.

This enables you to place both strengths and weaknesses in order of importance.

3. **Opportunities and threats**

Look at the organisation's external environment and ask what is happening (or might happen) outside the organisation which could:

- provide the chance of a new/revised course of action possibly beneficial to the organisation?
- seriously hamper the organisation's ability to compete/threaten its existence?

This involves looking at e.g.

social trends
technological developments
economic trends
political changes, actual or potential
etc.

Resulting answers identify opportunities open to the organisation and threats faced, e.g.

new product development
financing schemes
etc.

Place both opportunities and threats in order of importance.

4. **The next stage**

Produce prioritised list of: strengths
weaknesses

opportunities
threats.

Consider how strengths might be built on, in order to:

take advantage of opportunities
mitigate threats.

Consider how weaknesses might be tackled, in order to:

take advantage of opportunities
mitigate threats.

SUPPLEMENTARY NOTE ON SWOT

Reasons for use of SWOT

1. Provides a rational framework beneficial for:
   - decision-making
   - understanding current position.
2. Enables a clearer understanding of key issues:
   - affecting organisation
   - needing to be addressed.
3. Helps to set standards:
   - relative to others
   - relative to competitors (best practice?).
4. Provides a checklist and thought stimulation for auditing:
   - internally
   - externally.

5 Provides a base for prioritising issues to be addressed.

6. Provides a base for identifying/focusing on potential future strategies.

A cautionary note

The above depends on the perspective of the beholder: it is useful to look at several staff members' perceptions.

Issues do not remain static: they change over time and depend on context.

Organisational considerations

The organisational context

SOCIAL	POLITICAL
TECHNICAL	ECONOMIC

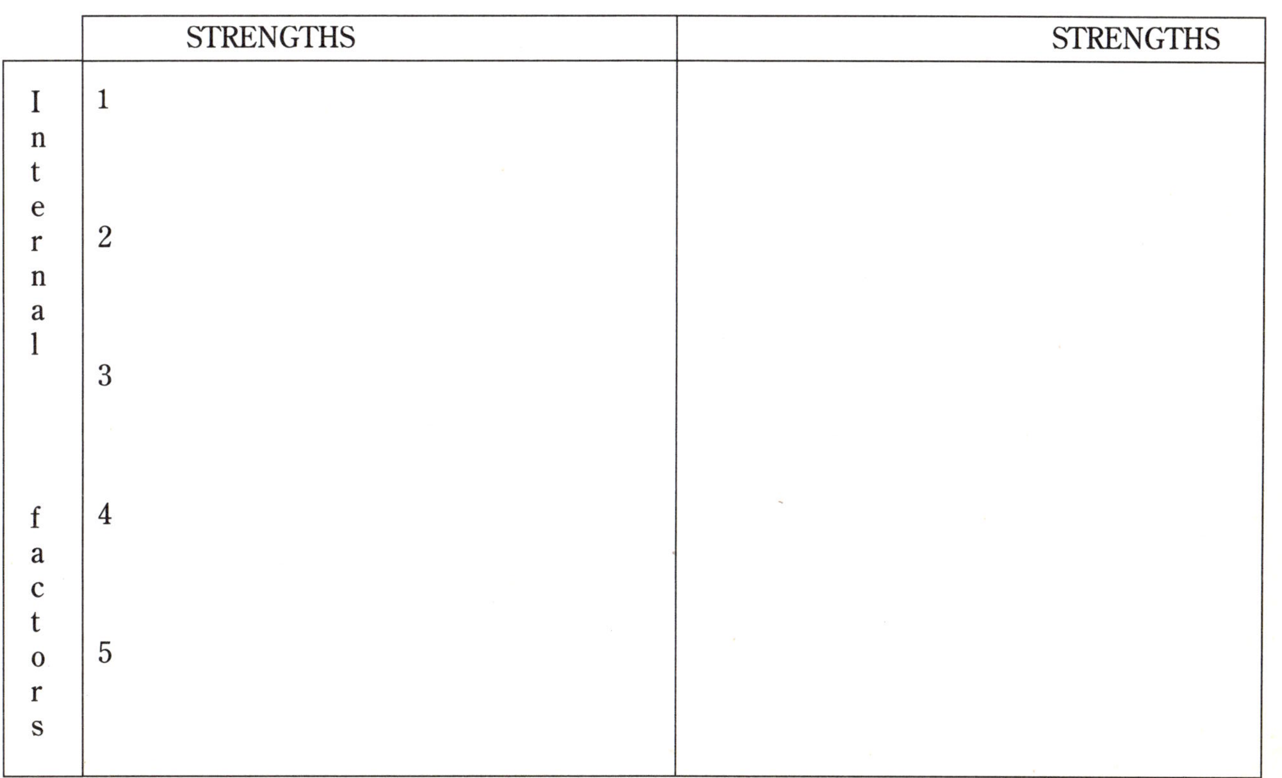

External factors

Internal factors	STRENGTHS	STRENGTHS
	1	
	2	
	3	
	4	
	5	

External factors

	OPPORTUNITIES	THREATS
1		
2		
3		
4		
5		

Appendix B: Cashflow pro forma

MONTH												
RECEIPTS	BUDGET	ACTUAL	BUDGET	ACTUAL	BUDGET	ACTUAL	BUDGET	ACTUAL	BUDGET	ACTUAL	BUDGET	ACTUAL
Cash Sales												
Cash from Debtors												
Capital Introduced												
TOTAL RECEIPTS (a)												
PAYMENTS												
Payment to Creditors												
Salaries/Wages												
Rent/Rates/Water												
Insurance												
Repairs/Renewals												
Heat/Light/Power												
Postages												
Printing/Stationery												
Transport												
Telephone												
Professional Fees												
Capital Payments												
Interest Charges												
Other												
V.A.T. payable (refund)												
TOTAL PAYMENTS (b)												
NET CASHFLOW (a-b)												
OPENING BANK BALANCE												
CLOSING BANK BALANCE												

Appendix C: Mission statement

We at XYZ want to be seen as a world-class company. This involves making our name synonymous with:

- Quality
- Reliability
- Total customer satisfaction.

Our mission to achieve excellence is based on five fundamental principles, as follows.

1. *Customer quality orientation* – customers define business; they need to be treated as the most important partners in the enterprise. Both sides (organisation and customers) will greatly benefit from success. Quality reputation wins customers. We want (and need) to be seen as reliable and caring towards our customers, and to understand their problems, requirements and needs. To be ahead of the competition, we have to anticipate our customers' needs (e.g. through innovation). To compete and win, we must redouble our efforts not only in the quality of our goods and services, but in the quality of our response to customers.
2. *Worker orientation* – success depends on involvement and empowerment of trained and highly motivated people. Empowerment is about giving people trust, respect and encouragement to use their creative potential; managers have to learn to lose a little power, workers to take on more power and personal responsibility for work and for

supporting the organisation's reputation and image. The workers have to be totally committed and take pride in their job. In gaining commitment, peer pressure (teams), involvement and pride are as influential as money. Workers must share in the achievement – through celebration of success as opposed to 'execution' for failure and mistakes.

3. *Community orientation* – involves commitment to a particular community (through the provision of employment and security), environmental issues and interaction with the community, as well as wider-scale considerations of national pride.
4. *Profit* – when it comes to world-class excellence, profit usually comes last. However, most league tables, comparisons, *Fortune 500* lists or other indicators of success are expressed in financial terms. Sometimes, however, these indicators may be misleading as far as quality is concerned. But if we do not make a profit, we disappear. Increased profitability is both our security and our guarantee. Profit is seen as a natural by-product of doing something well, not an end in itself.
5. *Growth* – this criterion relates not just to the physical growth of XYZ, but also to our ability to cultivate our strength to maintain a leading position and our ability to change, and change fast. Diversification, if done willy-nilly, does not pay. However, a certain amount of diversification is necessary to respond to market needs and maintain flexibility.

To work this way, XYZ management will:

- provide good leadership and coaches, with clearly expressed and understood values

- ensure that meeting the needs of customers is the guiding philosophy for the whole organisation
- be motivated by considerations of the organisation's long-term position rather than short-term financial performance; market position/share is more important than immediate profitability
- be hungry for market information
- be swift and responsive in handling complaints
- put considerable emphasis on quality control
- innovate constantly
- emphasise production/service quality rather than price
- listen to its employees, customers and competitors
- work on facts, testing and benchmarking
- maintain a culture that values continuous improvement and learning.

Appendix D: Employment contract checklist

The elements of the employment contract include:

- names of the employer and the employee (i.e. the remote worker)
- date when employment starts
- a statement whether any employment with a previous employer counts as part of the employee's continuous period of employment, and if so, when it began, or the fixed period of the contract
- rate of pay
- payment interval (per project, per day, per hour, etc.)
- hours of work
- holiday pay (if agreed), including entitlement on termination
- sick pay (if agreed)
- notice required to terminate employment from both the employer and the employee
- job title
- full description of responsibilities
- employee benefits (equipment provision or loan, any payments related to the equipment provision or use, insurance or any other relevant benefits)
- references to disciplinary rules and procedures, grievance procedures.

The contractual documents may also include the letter of appointment, the staff manual and work rules.

Appendix E: Payment issues

Organisations may deal with various types of remote worker and have different contractual arrangements with each type. Payment systems will be largely dependent on this contractual relationship.

For workers who are not the direct employees of the company payment will usually be based on results – i.e. an agreed output delivered at an agreed time to an agreed quality. This seemingly simple formula is, of course, dependent on clearly establishing the expectations of both parties before the contract is agreed, setting out a timetable and monitoring progress throughout the contract period.

Where the remote worker is a direct employee of the company, payment can often be seen as a difficult issue; however, with some careful planning and good management practice, problems can be minimised. Often a difficulty is a perceived loss of control. Some managers feel uncomfortable with not being able to see staff perform their work in situ and, for that reason, feel they cannot be sure of receiving value for money. If a job is easily quantifiable or has measurable outputs, the fears can be allayed. Many organisations have well-established norms by which the performance of an individual can be judged. Typing is a good example of this, for it will be relatively easy to set productivity standards for typists who work remotely. Performance can be monitored on a systematic basis and value for money judged accordingly.

Performance targets should be set carefully. If the targets are too difficult to attain, then the worker may become demotivated and resentful; if targets are too easy, complacency may set in. A flexible attitude where both parties can contribute to the review of targets is desirable.

With professional staff whose work is harder to quantify, the key to success is in good communications, setting targets, monitoring progress and providing feedback (see Chapter 6 on Managing Remote Working, and Chapter 7 on Project Management).

As a general rule, remote workers who are direct employees should be included within the pay structure of the organisation. This will help remote workers to feel that they belong and have not been marginalised as a result of their remote worker status.

Recommended reading

Barrow, C. and Barrow, P. (1992), *The Business Plan Workbook*, London: Kogan Page.

Barrow, L. (1988), *Financial Management for the Small Business*, 2nd ed., London: Kogan Page.

Bird, D. (1989), *Common Sense Marketing*, London: Kogan Page.

Bird, M. (1990), *Making Profit: A Six Month Plan*, London: Piatkus.

Blackwell, E. (1989), *How to Prepare a Business Plan*, London: Kogan Page.

Brooks, M.J. (1986), *Sources of Free Business Information*, London: Kogan Page.

Dickey, T. (1992), *Budgeting: A Practical Guide for Better Business Planning*, London: Kogan Page.

Finch, B. (1992), *Business Plans: 25 Ways to Get Yours Taken Seriously*, London: Kogan Page.

Hague, P. and Jackson, P. (1987), *Do Your Own Market Research*, London: Kogan Page.

Harrison, T. (1987), *Handbook of Advertising Techniques*, London: Kogan Page.

Mott, G. (1990), *Accounting for Non-accountants: A Manual for Managers and Students*, 3rd ed., London: Kogan Page.

Northern Regional Management Centre (1994), *Helpcard 4: Business Planning*, Management Helpcards, Washington, Tyne & Wear: NRMC.

Sterret, P. (1992), *Marketing Ideas for the Small Business*, London: Mercury.

Turner, S. (1987), *Practical Sponsorship*, London: Kogan Page.

West, A. (1991), *A Business Plan*, London: Pitman.

Remember too that banks and Training and Enterprise Councils (TECS) provide free information and information packs for new businesses.

3 Human Resources

- Introduction
- Remote workers
- Recruitment checklist
- Selection
- Checking references
- Cost of recruitment
- Other cost issues – reduction of staff turnover and recruitment
- Training and development
- Appraisal
- Forms of training
- Training records

Appendices:

A. Job Description Form

B. TNL Application Form

C. Job Preview Checklist

INTRODUCTION

Chapter 2 explained how important it is to define the forms of relationship between remote workers and the management. Good business, however, never relies solely on formal contracts. Human relations can only be regulated so far.

In managing remote workers, all the usual issues of people management apply. Chapter 3 highlights the issues specific to the right selection of remote workers and to their training and gives examples of relevant good practice.

REMOTE WORKERS

Not everyone finds remote working acceptable. Typically, good remote workers have most (or all) of the following characteristics:

- maturity – a worker who needs constant supervision is not suitable
- self-discipline
- considerable job experience
- good communication skills (including new ways of communicating), overcoming shyness and asking for help or advice
- ability to establish a schedule – making and keeping consistent and productive work habits
- confidence in one's own abilities
- a highly organised worker
- a volunteer for the job
- self-motivated and dedicated
- independent
- flexible – can change work patterns day by day
- ability to work on one's own
- adaptability
- trustworthiness
- professionalism with clients
- good time management skills.

The special demands placed on the remote worker include:

- working alone for extended periods of time
- working without easy access to other employees, managers, resource materials or clerical support
- working without the time management discipline and planning associated with office work
- working different hours from colleagues
- working with (potential) sources of interruption and distraction – family, neighbours, TV or stereo, delivery people, repairmen, etc.

- lack of supervision

Potential remote workers include:

- full-time employees of the organisation who wish to alter their working arrangements
- carers (of children or the elderly)
- people with short-term or permanent disabilities who cannot commute or have no access to reliable public transport
- people who will trade off higher salaries/benefits for flexibility and autonomy in more or less the same job.

The above characteristics and demands need to be considered in selecting remote workers and can be used as a checklist for identifying potential problems, as well as training and development needs.

Management needs to be aware of the specific requirements of remote workers, as does the remote worker.

RECRUITMENT CHECKLIST

Before appointing someone, carefully consider the following points.

1. *Is there a job to be done?* Observe the markets and their demands. This may sound strange, but a not uncommon trap is to look for an immediate replacement when a person leaves. You may find instead that the work can be re-distributed among existing workers. One of the common reasons for setting up a remote working operation is to have a pool of workers available to respond to the peaks and troughs of work demands. You may have to develop a strategy to build

up this pool by recruiting new workers, even when there is no immediate work available.

2. *Identify the job*. This usually involves doing a job description and analysis and then determining the person specification (for sample forms see Appendix A to this chapter). The job analysis requires you to identify the:

   - *purpose* – to what objectives does it contribute; what is its main purpose?
   - *tasks* – the key elements of the job.
   - *results* – the key outcomes.
   - *measures* – of performance.
   - *standards* – what is good or poor performance; what performance is needed?

   After analysing the job, draw up the job description under the following headings:

   - Title
   - Position in the organisation
   - Key tasks
   - Key results
   - Resources involved
   - Conditions (location, times, hours, etc.)
   - Salary/wages
   - Other benefits.

   You can now draw up a candidate profile:

   - skills needed
   - competences
   - personal competences
   - aptitudes

- general intelligence
- attainment
- disposition
- motivation
- circumstances
- interests.

3. *Look for the right person.* Check sources of recruitment, including:

   - internal recruits
   - employment agencies
   - job centres
   - professional bodies
   - advertisements
   - personal contact
   - head-hunting
   - colleges and universities
   - word-of-mouth recommendations.

   Remember to be flexible about candidates' possible restraints such as dependants, transport, and so on.

   The right method of recruiting is also important; it must not deter minorities and the disabled.

4. *Advertise.* Advertising should specify:

   - what on-site attendance is expected of the remote worker (if any)
   - any necessary visits to clients
   - the expected employment status of the remote worker
   - duration of the contract.

5. *Application form.* See Appendix B to this chapter for an example.

6. *Decide on the right person.* Check information/references, shortlist candidates and carry out selection procedures.
7. *Administration: person appointed.* Draw up contract of appointment, induction procedures and training requirements.

SELECTION

Selection criteria are based on the person specification which emerges from the job specification and are an objective means in the selection process.

General guidelines on selection criteria are the following:

- A person who has spent more time doing a particular job will do better than a person who is new and still learning the basics.
- A person with above-average performance in the current or previous job will do better than an average or below-average worker.
- A person with a broad background, with experience of a variety of assignments, will do better than a person with experience of fewer kinds of tasks.

In addition to the general characteristics of a good remote worker given earlier (pp. 44–45), specific requirements for remote workers include:

- proven track-record – if it is an employee (an internal recruit), with a minimum of three months' experience on-site; if it is an external recruit, with a minimum of three years' experience in a given job (the duration can be adjusted at the discretion of the employer).

- ability to demonstrate technical skills (i.e. job skills = can do the job).
- it is best to avoid a person who is completely new to the idea of remote work (i.e. to whom it just appeals).
- ability to demonstrate technological skills (i.e. ability to use equipment necessary for remote working – computer, modem, fax, etc.).
- suitability for remote working-based on profile of the remote worker.

To establish the suitability for remote working, the following methods can be used.

1. *Job preview* – enables the potential remote worker to meet existing workers. (This is a good way to assess one's own suitability for this kind of work. How does it feel to work alone? How does one motivate and discipline oneself? What are the practical problems?) This can be expanded into a framework of questions to give to the potential worker to discuss: see Appendix C to this chapter. This should help the candidate make an informed decision on whether to continue.
2. *Structured checklist of logistical issues* – checking, for instance, suitability of the candidate's workplace (i.e. home), arrangements for dependants that might interfere with the way of the working. These considerations can be checked on the candidate's say so, or on a visit to his/her home (see the example in Appendix D).
3. *Interviews* – by using the Application Form (see Appendix B to this chapter); the Job Preview Checklist (Appendix C); the Checklist of Logistical Issues (Appendix D); and Skills Form (Appendix E). Additional information should be obtained on:

- the candidate's perception of his/her work habits
- satisfactions/dissatisfactions or frustrations on the job
- views on the supervisor and co-workers
- role of work in the candidate's life.

Here the information you want is:

- how much the candidate depends on contact with others in the office
- how comfortable he/she is with different kinds of supervision
- what his/her expectations are about remote working.

The forms in the Appendices are a means of identifying essential information. The interview itself should be used as an opportunity to:

- check the facts
- probe and clarify any claims
- begin the development of personal relationship.

A checklist of good practice is given in Appendix F to this chapter.

4. *Personality inventories* – there are still no instruments on the market specifically designed to measure suitability for work at home; in this absence, the *Work Environment Scale*, *Myers-Briggs Type Indicator* and the *Work Style Reference Inventory* may be used. These instruments do not directly indicate if someone will do well as a remote worker, but provide information about traits and characteristics which seem to be essential for successful remote working, namely the level of need for contact with others, strength of preference for certain kinds of work and self-discipline. These tools usually require the collection of data over time to

make statistical analysis worthwhile. They need to be used with caution and may require expert guidance; the address of the British Psychological Society is: 48 Princes Road East, Leicester LE1 7DR.

5. *Trial work at home* – this could take place in the later stages of the selection process. It might be practicable to give potential remote workers pieces of work – well-defined and probably low risk – to assess their abilities. A decision will have to be made on whether this work will be paid or not.
6. *Work at home simulation* (as in 'games and simulations') – it can take the form of a group meeting: a presentation on remote working, with a video preview, if available. The candidates can exchange views and question the nature of remote work and observe the process in a 'safe' setting.

Additional sources of information on candidates include:

- current or previous manager – to indicate the candidate's working qualities, training level and the type of supervision required
- performance appraisals and other records.

When recruiting people who are already remote workers or have had experience, an alternative pro forma could be used.

You must design your form to suit the situation. For an example of an alternative form see Appendix G to this chapter.

CHECKING REFERENCES

References are mainly useful for confirming facts; they have less value in the forming of opinions or making judgements.

When checking a candidate's references, it is worth remembering to extract the following information:

- employment dates
- initial job responsibilities
- final job responsibilities
- supervisory needs
- team member performance
- conflicts encountered
- attendance/absenteeism
- strengths
- weaknesses
- learning capability
- record compared with others
- reasons for leaving
- how the employee is to be replaced
- the parting salary
- is the employee re-employable?

In checking up on the CVs, note:

- gaps in employment (beware a history that shows no gaps!)
- colleges cited, but no mention of degrees earned
- amazing educational backgrounds
- inflated achievements.

It may be useful to write to referees outlining the prospective work to be carried out. Be cautious how you interpret replies; you may need to read between the lines.

COST OF RECRUITMENT

Direct and indirect costs of human resource turnover include the following.

1. **Direct costs**:
   - employment advertising
   - agency and search fees
   - internal referrals
   - applicant expenses
   - travel expenses.
2. **Indirect costs**:
   - cost of management time per hire
   - cost of other employees' time per hire
   - cost of facilities per hire
   - cost of the learning curve and productivity business losses.

Good recruitment and *retention* strategies will cut the costs.

OTHER COST ISSUES – REDUCTION OF STAFF TURNOVER AND RECRUITMENT

An employer considering setting up a remote working organisation, whether through transferring existing staff or creating a greenfield venture, needs to consider the following issues.

1. *Child care* – formal child care arrangements need to be

made since, although the parent stays at home, she/he has to separate out work and child care. If this does not happen, then both children and work will suffer. The reduction of costs lies in the number of hours gained (no commuting, etc.), with fewer hours spent in a crèche or with a childminder/nanny. This also gives the remote worker an option and the opportunity to work partly at home and partly outside, together with reduced child care costs.

2. *Disability* – it may be wise to retain an existing worker who becomes disabled. The disability may, of course, be a temporary one (e.g. recovery from surgery or an accident) or permanent. However, if the worker is still capable of carrying out work duties, it is easier to retain him/her. But it must always be remembered to have the worker's health in mind – not to overstretch him/her, but perhaps redesign the job.
3. *Relocation* – costs here can be removed altogether (note, however, that you need to meet, so be careful of travel costs and time).
4. *Personal preference* – some employees are square pegs who do not fit easily into the organisation, or do not get on with colleagues; yet they may be skilled workers who should be allowed to continue with their employment.
5. *Retirees* – often highly skilled people, with long experience on the job, can be retained or re-employed, usually on a part-time basis, to best utilise their expertise.

TRAINING AND DEVELOPMENT

Every company should have a **staff development system** (for an example of one for regular employees see Appendix H to this chapter).

Training and development of remote workers, however, is an area that is often swept under the carpet and left to the discretion of the employing organisation.

Remote workers miss out on the ongoing informal training that takes place in an office situation – picking up tips and advice, asking colleagues for help, and so on. Because of the lack of constant physical contact with remote workers, it is often difficult to establish their specific training needs. Remote workers themselves also complain that they are not informed of training opportunities available to them.

There is also the (often unresolved) problem of who should pay for the training, that is time spent on training rather than doing the actual work – especially important in the case of workers paid per hour; and then the costs of the actual training itself (including travel, materials, etc.).

There is an additional need for increased communication with the remote worker during training, to provide psychological as well as technical support.

Training needs can be identified (see Appendix I to this chapter):

- at the selection interview and induction
- from observation of work performance
- through formal appraisal.

Induction training for remote workers should include the following:

- general information about the organisation
- the forms of communication to be used, especially computer-based ones

- information about regulations (health and safety, reporting systems, etc.)
- information on benefits and entitlements
- specific information about the role within the organisation, hierarchy, key support people, technical support and sources of help
- self-management
- time management
- planning/priority setting
- coping with stress
- telephone skills
- interpersonal skills
- keeping up to date with technological developments that may affect the remote worker.

For an example of an induction pack see Appendix J to this chapter. In the course of work, management will be monitoring work performance. How this is done is explained in Chapters 6 and 7, where it will be seen that formal systems are critical to the success of a remote working operation.

APPRAISAL

Appraisal usually means a formal process whereby management and workers meet to discuss performance and agree improvements and developments and also to review objectives.

The aim of any form of appraisal is to see how well the worker is doing and how he/she can improve his/her performance. Without appraisal, average (and below) performance is not corrected and above-average performance is not rewarded.

Appraisal is useful in:

- identification of training and development needs
- setting performance objectives (agree targets for the future, a chance for workers clearly to see what is expected of them; and see how workers' own objectives fit with the objectives of the business)
- motivating workers (employees get feedback on their performance; see how well they are doing against set objectives; and feel in control of their work)
- rewarding performance (recognition and praise in the form of benefits; reinforcements of desirable performance; and feedback on positive aspects of performance).

Informal appraisal happens all the time in an on-site situation through everyday contact. It provides encouragement and regular feedback on performance, reinforces positive performance, corrects undesirable performance and also reviews a short period of time.

In the case of remote working, the only chance of informal appraisal occurs during any meeting with the remote worker, or through the formal reporting system. In fact it should be built in as a review of:

- work progress
- the remote worker's feelings concerning his/her involvement in a particular project
- levels of performance
- expectations
- problems and difficulties that both remote worker and manager feel need voicing.

Formal appraisal may happen just once or twice a year. An

example form is given in Appendix I to this chapter. Appraisal needs careful preparation; it reviews longer periods of time and is often linked to pay reviews.

Frequency of formal appraisal is usually left at the discretion of the individuals involved; however, the following points in time should be taken into consideration:

- beginning of a project (job, task)
- deadlines as established before work starts
- end of project
- progress review.

For formal appraisal dates have to be set, in order for both sides to prepare. Workers should be asked to think about their performance up to date, in the past six months, or since the last appraisal, as well as their goals and targets for the next six months, next projects, and so on.

A time and place should be agreed to carry out the appraisal interview. During the interview, the worker needs to be asked how he/she feels about their work: Are there any problems, worries or difficulties? Three actions need to take place: listening, discussion and agreement. Past performance needs to be reviewed against previously agreed targets, also gaps in the performance and the reasons for them. Discussion will concern training and development needs. New targets need to be agreed, together with any training and development plans.

The following are the main features of appraisal:

- *trust and respect* – the worker must not feel defensive or that he/she is being criticised; accurate knowledge of performance helps the dialogue.
- *common standards* – the same criteria have to be used for

everyone (no easier targets for some and harder for others); these standards have to be measurable and realistic.

- *dialogue* – workers have to be able to voice their views, concerns and suggestions for improvement of performance.
- *feedback* – this should be accurate, constructive and regular.
- *follow-up* – should happen as soon as possible after a decision has been made to see if action has been taken.

Formal appraisal is easier to manage when there is an employer–employee relationship. Where the remote worker is self-employed, formal appraisal is still possible, though the approach may differ. In each appraisal there is a balance to be drawn along the spectrum:

Manager evaluates worker ◄——► Worker self-evaluates

If the appraisal is mainly to the left of the spectrum, there is a risk the worker will feel adjudged and criticised. This may be accepted practice in organisations where management has some formal position of power over the worker, though it is not generally considered to produce the best results.

In remote working, especially where the remote worker is self-employed, it is best that the appraisal is mostly to the right of the spectrum. This requires the manager to adopt a less judgemental and more counselling approach. For examples of good appraisal questions see Appendix K to this chapter.

In remote work supervision, appraisal is time-consuming and complex and carries with it additional costs (mainly management time).

FORMS OF TRAINING

Training is often described as a **training intervention** – i.e. a conscious decision is taken to intervene in a situation to improve what is being done. All training should be conscious and planned.

Training can take the forms of:

- courses and programmes at a college or at home
- on-the-job instruction, by a specialist trainer or by the manager
- coaching.

Remote workers will find many courses are available for study at home, in personal skills and technical skills, and both paper-based and IT-based. (Rather than attempt to go into detail here, it is worth contacting local colleges and TECs for advice.)

Courses and programmes provide a good opportunity for remote workers to get out of the house and to meet other people, often in a similar situation. However, regular attendance may be difficult because of problems with transport or care of dependants.

On-the-job instruction can only be done face to face, so if management determines a remote worker needs to be shown how to do a specific job, then meetings will have to be arranged.

In the case of courses and programmes, the contract between the management and the remote worker will need to determine who pays for what. In practice, a judgement has to be made by both parties since not all training issues may be black and white. If the training intervention benefits both parties, then it would be reasonable to negotiate sharing of costs. If training is offered by management as a means to attract potential remote workers, then management might judge this to be a reasonable part of recruitment costs.

It is in management's best interest to encourage remote workers to continue their learning and development recognising, however, that remote workers have certain rights, including the right not to develop! This is best done by management adopting a coaching style; in particular, management should:

- look for opportunities to provide feedback
- look for work tasks that stretch the remote worker, mixing routine tasks with the developmental ones
- adopt an allocation of work style that encourages the remote worker to say how best the work could be done, to identify potential problems and to offer contingency plans.

Where and how coaching takes place depends on the communication patterns of the business; but all communication channels offer the opportunity to coach, if management is prepared to look. Even e-mail discussions provide such opportunities (see Chapter 5).

Central to good coaching, or indeed any training, is reflection. Much of the effort put in by management will be in using questions that encourage remote workers to *think through* for themselves: how well have they performed and what can be done differently and better? (The approach is similar to that described in the section on 'Appraisal'; see also Appendix K to this chapter.)

Remote workers need feedback; they need coaching; and they need nurturing. It is all too easy to forget this when workers are at a distance and not seen very often.

TRAINING RECORDS

Management will need to keep records – especially if the pool of remote workers is large. These need to be collated into a group record, so that management can see as a whole, what skills and experience are available. For two sample forms see Appendix L to this chapter.

It is strongly suggested that extracts from these records, and from the recruitment records, are put into electronic format as 'pen-portraits' for all remote workers (or those working together on a particular project) to see.

Appendix A: Job Description Form

JOB DESCRIPTION

Title:

Position in the organisation/Relationship with others:

Key tasks:

Key results:

Resources involved:

Conditions (location, times, hours, etc.):

Salary/Wages:

Other benefits:

JOB DESCRIPTION – WHAT IS INVOLVED

Set out below is the information that should be included.

Job title.
Relationship with others – reporting, supervision, liaison, coordination and the volume of contact needed and with whom.
Job content – tasks and duties, including important and intermittent tasks; the analysis of the main tasks gives a good picture of the job in its present form, in order to assess its suitability for remote working. In the case of project work, look for ways of segmenting it. Progress can then be monitored more easily.
Working conditions – physical and social environment, times of work, salary and benefits. These may include e.g. acceptable levels of noise and lighting, as well as health and safety, legal, and tax and insurance aspects of the job.
Performance standards – both quantitative and qualitative (according to the type of job).
Human requirements – physical and psychological characteristics of persons suitable to do the job.

JOB ANALYSIS

Purpose:

Tasks:

Results:

Measures:

Standards:

CANDIDATE PROFILE

Needed skills:

Needed competencies:

Needed personal competencies:

Aptitudes:

General intelligence:

Attainment:

Experience:

Disposition:

Motivation:

Circumstances:

Interests:

Appendix B: TNL Application Form

1. Name: ....

2. Address: ....

....

....

3. Home telephone no.: ....

4. Work telephone no.: ....

5. Date of birth: ....

6. Nationality: ....

7. Marital status: ....

8. Do you belong to an ethnic minority? ☐ **Yes** ☐ **No**

   If yes, please state which:

   ....

9. Do you have a disability? ☐ **Yes** ☐ **No**

   If yes, please state what kind:

   ....

10. Available for work . . . ?

    – full-time ☐ please specify days: ....

    – part-time ☐ and/or no. of hours: ....

11. Date available for work: ....

12. Education and qualifications:

School/college/ university	Qualifications	Grade	Date

13. What is your present employment status?

Employed: ☐ Full-time ☐ Part-time ☐

Unemployed: ☐ For how long? ....

Self-employed: ☐ For how long? ....

14. Are you familiar with your tax status and possible benefits available to you?

Yes ☐ **No** ☐

15. Previous employment to date (not just in IT):

Name of employer	**Nature of work/position**	**Time spent there**	**Reason for leaving**

16. Please list other experiences, skills and qualifications:

Experience	**Skills**	**Qualifications**	**Date**

17. Can you use?

	Yes	**No**	*(please tick)*
a computer	☐	☐	
a fax	☐	☐	
a modem	☐	☐	
a spreadsheet	☐	☐	
a word processor	☐	☐	
an electronic bulletin board	☐	☐	
e-mail	☐	☐	
a web browser	☐	☐	

18. Which of the following do you have in your home and can use for work?

	Yes	**No**	*(please tick)*
a separate telephone line	☐	☐	
a PC	☐	☐	
a printer	☐	☐	
a fax	☐	☐	
an answering machine	☐	☐	
a modem	☐	☐	
a photocopier	☐	☐	
a notebook computer	☐	☐	
a cellphone	☐	☐	
An Internet connection	☐	☐	

19. *References*: Please name 2 persons and their addresses with telephone numbers who we can contact for references:

Name:	....	Name:	....
Address:	....	Address:	....
	....		....
	....		....
	....		....
	....		....
Phone no.:	....	Phone no.:	....

Appendix C: Job Preview Checklist

1. Has the candidate worked from home before?
2. How does the candidate feel about working alone?
3. Can the candidate separate professional work physically (i.e. separate room) and psychologically/emotionally from the otherwise near proximity of home?
4. Is the candidate likely to interrupt work him/herself?
5. Does the candidate have many external interruptions?
6. How does the candidate manage interruptions?
7. Does the candidate take a lot of breaks during the working day?
8. How does the candidate organise his/her working day?
9. Does the candidate know exactly what he/she will be doing the following day?
10. What are the candidate's practical problems in working from home?
11. What solutions has the candidate for overcoming these problems?
12. Does the candidate have home/family commitments that may interfere with his/her work?
13. What is the motivating factor in the candidate's work (i.e. employment, income, anything else)?

Appendix D: Checklist of Logistical Issues

1. Does the candidate have a separate room or space solely devoted to his/her work?
2. Is it adequately equipped, furnished, lit and heated?
3. Who does the candidate share his/her working space with?
4. If the candidate were to increase the volume of equipment or furniture necessary for his/her work, would the workspace still be satisfactory?
5. Do members of the candidate's family have easy access to his/her workspace?
6. Does the candidate work outside office/school hours?
7. If the candidate works outside office/school hours, does home life interfere with the candidate's work?
8. Do members of the candidate's family help him/her with work? If so, in what way?
9. Does the candidate have a computer for exclusive use?
10. Does the candidate have a separate telephone line exclusively for work?
11. Is there anyone in the candidate's household who stays at home during his/her working hours (assuming 9–5) – children, spouse or elderly relatives?
12. What arrangements has the candidate made (or proposes to make) for the care of dependants?
13. Does the candidate have to pay for care of dependants?
14. Does the candidate own a car?
15. If the candidate does not own a car, is public transport in his/her area satisfactory?

Appendix E: TNL Skills Form

The information gathered from this form will help us create a skills matrix. It is also the basis for the recruitment interview.

I. TECHNICAL SKILLS

1. Primary job functions – software

   Which of the following do you have experience in? (*please tick all applicable*)

   General management ☐

   Project management ☐

   Systems engineering ☐

   Business analysis ☐

   Systems analysis/design ☐

   Team leading ☐

   Programming ☐

   Technical support ☐

   Other ☐

   (*please specify*)

   ....

2. Please list the computer equipment you have used. Specify below if Mainframe(Ma), Mini (Mi), Workstation (W) or PC:

Make	Model	Ma	Mi	W	PC	No. of years	Function

3 Which systems software have you used and how do you rate your knowledge? (*please tick*)

System	Knowledge		
	Some	Considerable	Expert
Windows			
DOS			
UNIX			
Other			

4 Which programme languages have you used and how do you rate your knowledge? (*please tick*)

Language	**Knowledge**		
	Some	**Considerable**	**Expert**
Visual Basic			
Delphi			
C			
C^{++}			
COBOL			
PARADOX			
SQL			
HTML			
Other			

5. General business experience – do you have any experience of the following? (*please tick all applicable*)

Administration	☐	NI	☐
Banking	☐	Stock control	☐
Tax	☐	Management	☐
Financial management	☐	Sales	☐
Training/education	☐	Bookkeeping	☐
Legal	☐	Editing	☐
Marketing	☐	Word processing	☐
Research	☐	Data entry	☐
Graphics	☐	Government	☐
Communication	☐	Utilities	☐
Engineering	☐	Manufacturing	☐
Other (*please specify*)	☐		

6. Are there any areas in which you consider yourself an expert?

Yes ☐ No ☐ (*if yes, please specify*):

....

II. PERSONAL SKILLS

1. Have you worked from home before? Yes ☐ No ☐
   If yes, please specify the length of time and nature of work:

2. Did you find it easy to work in this way? Yes ☐ No ☐
3. What were the problems you encountered?

   ....

   ....

4. How did you overcome them?

   ....

   ....

5. On average, how long can you work for without breaks?

   ....

6. How frequent are your breaks?

   ....

7. How do you cope with distractions?

   ....

8. How important to you is frequent and regular contact with ... ? (*please tick*)

	Not important	**Fairly important**	**Very important**
Colleagues			
Boss/project manager			
Outside world			

9. How frequently would you expect to communicate with ... ? (*please specify*)

Colleagues ....

Boss/project manager ....

Outside world ....

10. Please mark in order of preference (*1 – prefer, 5 – prefer not to*) means of communication you would use with colleagues and bosses/project managers:

in person ....

by phone ....

via e-mail ....

by letter ....

III. PROCESS SKILLS

1. Have you ever kept detailed time sheets/work diaries or logs or any other work records before?

   Yes ☐ No ☐

2. Would you say your report writing skills are … ?

   average ☐

   good ☐

   above average ☐

3. Have you used electronic means of communicating before (e.g. e-mail, bulletin boards, video conferencing?)

   Yes ☐ No ☐

IV. TEAMWORKING SKILLS

1. Have you worked as a member of a (project) team before?

   Yes ☐ No ☐

2. How would you rate your participation in team work?

   poor ☐

   average ☐

   good ☐

   very good ☐

V. RELATIONSHIP WITH CUSTOMERS

1. Have you ever had to deal directly with customers?

   Yes ☐ No ☐

   If yes, were you required to visit the customers or work from their premises?

   Yes ☐ No ☐

2. How frequent was the contact with the customers?

Appendix F: Interviewing

An interview is a *conversation with a purpose* – the purpose being to obtain as much information about a person as possible and to assess his/her suitability for a job.

The interview is structured to extract information on the candidate, which will enable a full assessment of his/her experience, qualifications and personal qualities to be made. This will then be measured against the requirements of the job specification.

The sequence of an interview is as follows:

- Put the candidates at ease – he/she will be nervous and needs to be relaxed to put across their thoughts freely.
- Explain the stages of the interview.
- Briefly describe the job and what is involved, and also describe the company.
- Ask the candidate to outline his/her education and career up till now; concentrate on the most recent facts, and watch out for glossing over; and analyse what you hear critically, but with no prejudice (especially reasons for leaving previous jobs).
- Provide an opportunity for the candidate to talk about his/her achievements.
- Allow time for the candidate to ask questions.

DOS AND DON'TS OF INTERVIEWING

(After M. Armstrong, *How to Be an Even Better Manager.*)

Do:

- plan the interview
- establish an easy and informal relationship
- encourage the candidate to talk
- cover the ground as planned
- probe as necessary
- analyse career and interests to reveal strengths, weaknesses and patterns of behaviour
- maintain control over the direction and time taken by the interview.

Don't:

- start the interview unprepared
- plunge too quickly into demanding questions
- ask leading questions
- jump to conclusions on inadequate evidence
- pay too much attention to isolated strengths and weaknesses
- allow the candidate to gloss over important facts
- talk too much.

Appendix G: Remote Worker Questionnaire

1. Name: ....

2. Address: ....

....

3. Please specify your work situation before joining TNL:

— worked in a different job ☐

please specify ....

....

— worked in different location ☐
(i.e. have moved)

— there was no opportunity to carry out previous job because:

moving home ☐

redundancy ☐

pregnancy/small child ☐

4. When did you start working from home? ....

5. Please specify details of work carried out at home:

Type of work				
Amount of work				
Basis of contract				
Equipment used				

6. At present, do you do any work other than for TNL?

Yes ☐ No ☐

Is the work of the same or similar nature?

Yes ☐ No ☐

7. Please indicate which apply to you:

self-employed ☐

employee with a fixed number of hours ☐

piece-rate payment ☐

payment by the hour ☐

payment per project ☐

8. What equipment do you supply (i.e. pay for yourself)?

....

9. What equipment is provided for you?

....

10. Do you have a family? Yes ☐ No ☐

    If yes, how do they feel about your working arrangements?

    ....

    ....

11. Do your home circumstances influence your choice to be a remote worker? If so, in what way?

    ....

12. Do any members of your family help you in your work?

    Yes ☐ In what way? ....

    No ☐

13. How long do you usually work for?

    average working week ——— hrs (per week)
    longest working week ——— hrs (per week)
    average working day ——— hrs (per day)
    longest working day ——— hrs (per day)

14. Does the length of your working day vary greatly?

    Yes ☐

    No ☐

15. Do you normally work within normal working hours (9–5)?

Yes ☐

No ☐ Do you usually work ... ?

- evenings ☐
- nights ☐
- weekends ☐

16. Please state the amount of work done outside normal working hours (9–5):

none/very little ☐

about 1/3 ☐

about 1/2 ☐

about 3/4 ☐

almost all ☐

17. Have you been without work within the past 12 months (excluding planned time off)?

Yes ☐

No ☐

18. Would you say that generally you have … ?

    too little paid work ☐

    enough paid work ☐

    too much paid work ☐

19. Would you prefer … ?

    much less work ☐

    less work ☐

    about the same amount ☐

    more work ☐

    much more work ☐

Please give reasons: ……………………………………………………………………

20. Can you influence the amount of work you do?

    No ☐

    Yes, a little ☐

    Yes, considerably ☐

    Yes, completely ☐

21. What are, in your case, the major advantages and disadvantages of remote working?

	Advantage	*Neutral*	*Disadvantage*
commuting and travel	☐	☐	☐
close supervision	☐	☐	☐
frequent interruptions	☐	☐	☐
combining work and care	☐	☐	☐
working when it suits me	☐	☐	☐
meeting the family's demands	☐	☐	☐
amount of time spent with the family	☐	☐	☐
work time mixing with free time	☐	☐	☐
amount of working space	☐	☐	☐
clerical support	☐	☐	☐
other office support (copying, etc.)	☐	☐	☐
amount of self-discipline needed	☐	☐	☐
development of working skills	☐	☐	☐

	Advantage	*Neutral*	*Disadvantage*
development of business start-up skills	☐	☐	☐
status, as seen by others	☐	☐	☐
promotion chances	☐	☐	☐
benefits, perks, pension scheme	☐	☐	☐
level of pay	☐	☐	☐
self-employment status	☐	☐	☐
taking part in social events in the office	☐	☐	☐
taking part in working meetings	☐	☐	☐
contact with others in similar work	☐	☐	☐
other (*please specify:*)	☐	☐	☐

Appendix H: Staff Development System – sample format

INTRODUCTION

The purpose of this document is to remind you of the XYZ Staff Development System framework and to outline the rationale; and also to suggest ways in which you may like to implement the system to match your individual training and development needs.

It is XYZ's intention to use the proposed Development System as a developmental system and not an appraisal system.

XYZ's focus is on developing its people and on looking to present, future and personal training and development needs for roles within the company.

TRAINING AND DEVELOPMENT POLICY

The activities and approaches, and the broad future development plans, have recognised a need to respond to rapid changes. Global competition will mean an increasingly unforgiving environment.

XYZ has recognised that its people are its greatest assets, because only they can adapt and respond to changes. Quality performance will be critical.

This will only be enabled by training and development programmes for the individuals within XYZ. The training and development framework will encompass the Investors in People National Standard, and will be underpinned by procedures accredited to the BS 5750 Quality Assurance Standard and National Vocational Qualifications (NVQs). These three standards are not designed to take away the responsibility from the individual, but to aid individuals in training and development and in their present or future job role.

All staff will be aided in identifying their own needs for training and development in keeping with the business aims and objectives of XYZ. This will be achieved through XYZ's adaptable Development System.

PRINCIPLES

As stated previously, people are the most valuable assets of XYZ and the development of this important resource should aim to:

- maximise the contribution we each make to our job and the organisation
- maximise the satisfaction we each gain from doing our work.

These two aims are mutually supporting.

The main responsibility rests with the individual to identify training and development needs, and with the company to ensure these needs are met, both for the benefit of the individual and XYZ.

Effective development aims to meet both organisational and individual needs. It demands a continual dialogue which addresses some key questions:

'What is expected of me?'

'How am I doing?'
'How can I develop myself?'
'How can others help me to develop?'

In addition to this dialogue, every individual should have the opportunity of a more formalised development interview on a regular basis. XYZ's Development System addresses this need.

XYZ DEVELOPMENT SYSTEM FRAMEWORK

Job Description

The starting-point will be your own job description, drawn up by yourself. This can be based on a previously completed job description.

Skills and Knowledge Review

This will form the basis for you to complete a skills and knowledge review – *skills* being expertise, *knowledge* being understanding. The aim of this component is to establish *what you need* to be able to carry out your role in XYZ, and *what you already know* in order to do your job currently.

This will act as an aid in identifying your **personal development needs**. A relevant framework of Occupational Competence will be used. This will help you in highlighting any areas which need further development and strengths which should be built upon.

Personal Development Plan (PDP)

This will take into account factors such as XYZ's Position Statement and budgets. From this it will be decided what is required in terms of training and development, to enable you to meet XYZ's and your own goals and objectives.

This PDP will be set up on three equal levels of priority:

1. *Present* – will address what is required in terms of training and development for you to do your present job.
2. *Future* – will address training and development needs, taking into account XYZ's Position Statement and Future Direction and any other external factors.
3. *Personal* – will address training and development for personal development and personal aspirations.

The PDP form, along with the Development Agreement form, will be the only two forms in the 'formal' development system. This satisfies the requirement of the BS 5750 standard.

Development Agreement

To aid you in mapping out the direction of your training and development, the Development Agreement should be used to set out your aims and objectives for training and development: these may be based on many factors. With the XYZ Development Agreement, you help set the standards to be achieved. These need to be challenging and personally stretching, but remain achievable.

Development Interview

This is the stage where components of the Development System are reviewed and discussed, so that evaluation of where you are now can be established. The whole process (or parts of the Development System) can then go round the cycle again.

THE PROCESS

There are available several frameworks and models of Occupational Competences, which you may want to use or adapt for your individual needs. The choice is entirely yours.

The most important thing will be to answer the following questions:

'What is expected of me?'
'How am I doing?'
'How can I develop myself?'
'How can others help me to develop?'

Appendix I: Appraisal and Personal Development Form

NAME OF THE REMOTE WORKER: ....

DATE: ....

REVIEW

1. What jobs and/or tasks have been done satisfactorily since the last appraisal?
2. What learning and development objectives have been met since the last appraisal?

ANALYSIS

1. What are the organisation's future objectives that may require development by the worker?
2. What improvements on past performance are required of the worker by management?
3. What are the medium- and long-term aspirations of the worker?
4. What personal and job skills are not being fully utilised?

PLANNING

1. What personal and job skills development is needed? How will this be actioned over the next six months? What do the required improvements look like in practice?
   (*prioritise each – 1, 2, 3, etc.*)
2. Agree new work targets, defined in SMART (*S*pecific, *M*easurable, *A*chievable, *R*elevant, *T*ime-based) terms.

Appendix J: TNL Induction Pack

The contents of the TNL Induction Pack are:

TNL's Commitment to Quality
TNL's Guidance for Members
Requirements and expectations
Human Resource Development Policy
Remote worker communications – Computer Conference System (CCS)
E-mail etiquette

TNL'S COMMITMENT TO QUALITY – A POLICY STATEMENT

TNL supports a Quality Management System based on the principles of **Total Quality Management**. The system requires that all members of TNL accept responsibility for their own performance and ensure that their work is 'right first time'. The emphasis is on prevention, not detection.

In addition to satisfying the needs of the external customer, the quality requirements of internal customers (i.e. TNL management and/or other members of project team) must be observed.

In order to achieve commonality of practice for some types of work, the worker may be provided with guidelines and practical descriptions of TNL procedures. There will be instances when members will be expected to adopt mandatory standards. Undoubtedly, such standards may be seen by some as imposing undesirable constraints. However, it is our experience that a level of control is necessary to ensure efficient, effective and economical working practices.

The aim of TNL is to provide appropriate mechanisms which will

— Assist in the coordination and control of those activities necessary to achieve a desirable level of quality, at an acceptable cost.
— Enable procedures continually to be improved and to reflect experience gained, as well as changes in technology.

TNL'S GUIDANCE FOR MEMBERS – NEW WORK: THE SEQUENCE OF EVENTS

Step 1: Information supplied by TNL

When you are approached by TNL, you should be provided with as much information as possible. You will be asked to submit a proposal with an estimate for the job.

Step 2: The contract

Having agreed terms with the external customer, TNL will discuss your original estimate and consult you regarding any changes required.

The terms of your contract will be negotiated and the agreement will be formalised.

Step 3: Issuing the work

You will be provided with:

- job instructions/specifications
- general information about your roles and responsibilities, the project team set-up, etc.
- instructions as to which standards, guidelines and procedures are to be followed.

Step 4: Reviewing the work

You will be asked to participate in a review of your work, including:

- the overall success of the job
- a review of the processes
- an evaluation of your contribution to particular aspects of the job
- an evaluation of the relationship between yourself and TNL.

REQUIREMENTS AND EXPECTATIONS

When you take on a job for TNL, you will need to be clear about a number of operational issues. The following information should be provided.

1. **Individual roles, responsibilities and authorities**
   A written description of your own position with regard to work undertaken. This should state clearly your relationship with other TNL members and with TNL's client.
2. **Team members (when working as part of team)**
   A list of team members, their role within the project and details of how and when they can be contacted. This must include a clear statement of who should be approached regarding queries and problems.

3. **Resources**
   A list of the equipment and other resources required, with details of who is expected to provide it and/or how it will be made available.
4. **Standards, procedures and guidelines (for monitoring and controlling progress and quality)**
   Confirmation of TNL mandatory procedures applicable to the job (and notification of any common standards which do not apply), together with a list of other procedures which *must* be followed; and a detailed list of strategies and guidelines which should be referred to.
5. **Methods of communication**
   Details of which methods of communication should be employed for different purposes and their frequency; and the duration and location of meetings. Also rules regarding frequency and use.
6. **Administrative issues**
   A list of facilities available, together with procedures for submitting invoices, expense claims, etc.

HUMAN RESOURCE DEVELOPMENT (HRD) POLICY

TNL does not employ workers, rather it contracts with them. It cannot enforce a particular approach to worker development, but it does hope that both the worker and TNL can benefit if a careful and considered approach to HRD is adopted. To this end, TNL management will:

- Adopt a proactive approach to coaching workers, using normal day-to-day opportunities to provide constructive feedback. Depending on the time available and circumstances, a variety of coaching methods will be used – from **direct**

('You should have done ABC this way') to **indirect** ('How could you have done ABC differently?').

- Provide induction training to all new workers. This will include working through the company manual on work and remote working practices, company policies and philosophy, communication systems and terms and conditions; and will involve closer monitoring by management.
- Undertake a formal appraisal with each worker annually, to focus mainly on the worker's longer-term aims. This will be based on the worker's current skills and experience, likely demands from the company in the coming year, the worker's personal aspirations and development opportunities available.
- Undertake a formal appraisal with each worker at the end of each project, to review the worker's performance and the company systems, with the specific aim of improving the quality and quantity of the worker's contribution in the future.
- Identify for all workers what relevant training and development opportunities exist 'out there'.

TNL expects its workers, as self-employed professionals, to play a full part in their own development: workers are expected to be open and honest about their needs and realistic in their aims. They are expected to invest the time required to develop, or (at the very least) to invest the time to keep up to date in their areas of expertise.

REMOTE WORKER COMMUNICATIONS – COMPUTER CONFERENCING SYSTEM

We have established a **computer conferencing system** (CCS) as a core communication mechanism to link:

- remote worker and the company
- remote worker and remote worker
- remote worker and customer, where applicable.

The computer conferencing system is accessed by you from your workplace using your PC and a modem. You dial into it, 'logging on' and using the software provided. We expect you to dial in every day. You will use it

- to receive and send electronic mail
- to read and post messages for public consumption
- to read and send information concerning your projects.

The computer conferencing system is not designed to replace other forms of communication, particularly the use of telephone and face-to-face meetings. The value to the remote worker, however, is that the computer conferencing system is open 24 hours a day, so you can access stored information and leave messages at times that are convenient to you. It only works, though, if you keep to the disciplines required:

- dial in every day; get into the habit of checking your messages routinely
- be an active participant – and contribute as well as read, don't be passive
- be sensitive to other users – follow 'E-mail etiquette'.

Use of the computer conferencing system is an essential

requirement of the company. At the beginning of any particular project, you and management will agree:

- frequency of access to the system, and by whom
- information to be exchanged on every access
- how information on the project and worker activities will be stored.

Communications software

Details on the use of the communications software are given separately and are upgraded as the software is developed.

E-MAIL ETIQUETTE

(Summarised and amended from an Internet posting by C. V. Rospak, 25.04.93.)

There are some general principles that apply to all messages sent on the computer conferencing system. Some may seem obvious, but if we can make mistakes in normal communications when we can see or hear the other person, then electronic communications need to be handled with some extra thought.

- **Never forget that the person on the other side is human!**

Because your interaction with the CCS is through a computer it is easy to forget that there are people 'out there'. Situations arise where emotions erupt into a verbal free-for-all that can lead to hurt feelings.

Please remember that other people are reading your words. Do not attack people if you cannot persuade them with your presentation of the facts. Screaming, cursing, and abusing others only serves to make people think less of you and less willing to help you when you need it.

If you are upset at something or someone, wait until you have had a chance to calm down and think about it. A cup of (decaf!) coffee or a good night's sleep works wonders on your perspective. Hasty words create more problems than they solve. Try not to say anything to others you would not say to them in person in a room full of people.

- **Be careful what you say about others**

Remember that you read public postings, so do many other people. This group quite possibly includes your boss, your friend's boss and your partner's brother's best friend. Information posted on the CCS can come back to haunt you or the person you are talking about.

- **Be brief**

Never say in ten words what you can say in fewer. Say it succinctly and it will have a greater impact. Remember that the longer you make your message, the fewer people will bother to read it with the care it may deserve.

- **Your postings reflect upon you – be proud of them**

Most people on the CCS will know you only by what you say and how well you say it. They may someday be your co-workers or friends. Take time to make sure each posting is something that will not embarrass you later. Minimise your spelling errors and make sure that the message is easy to read and understand. Writing is an art and to do it well requires practice. Since much of how people will judge you on the CCS is based on your writing, this time is well spent.

- **Use descriptive titles**

The subject line of a message, 'the header', is there to enable a person with a limited amount of time to decide whether or not to read your message. Tell people what the message is about before they read it. A title like 'Car for sale' does not help as much as '66 MG Midget for sale'.

- **Be careful with humour and sarcasm**

Without the voice inflections and body language of personal communications, it is easy for a remark meant only to be funny to be misinterpreted. Subtle humour tends to get lost, so take steps to make sure that people realise you are trying to be funny! The CCS uses an idea borrowed from the Internet which has developed a symbol called the 'smiley face': it looks like ':-)' and points out sections of a message with humorous intent. No matter how broad the humour or satire, it is safer to remind people that you are being funny. You could use '(g)' for 'grin' instead.

Be aware that quite frequently satire is posted without any explicit indications. If a message has outraged you, then you should ask yourself if it may have been merely unmarked satire. Several self-proclaimed connoisseurs refuse to use smiley faces, so take heed or you may make a temporary fool of yourself; develop a thick skin!

- **Do not shout**

Using capital letters throughout is like SHOUTING. If you wish to emphasise something, use *asterisks* to highlight text.

- **Summarise what you are following up**

When you are following up someone's message, summarise the parts of the message to which you are responding. Use the quote facility. This allows readers to appreciate your comments rather than try to remember what the original message said.

Summarisation is best done by including appropriate quotes from the original message. Do not include the entire message since it will irritate the people who have already seen it. Even if you are responding to the entire message, summarise only the major points you are discussing.

- **Read all follow-ups and don't repeat what has already been said**

Before you submit a follow-up to a message, read the rest of the messages in the section to see whether someone has already said what you want to say. If someone has, don't repeat it.

- **Be careful about copyrights and licences**

Once something is posted on to the CCS, it is *probably* in the public domain unless you own the appropriate rights (most notably, if you wrote the item yourself) and have posted it with a valid copyright notice. A court would have to decide the specifics, and there are arguments for both sides of the issue. For all practical purposes, assume that you effectively give up the copyright if you don't put in a notice. Of course, the *information* becomes public, so you mustn't post trade secrets that way.

You should be aware that posting movie reviews, song lyrics or anything else published under a copyright could cause you, your company or members of the net community to be held liable for damages, so caution in using this material is highly recommended.

- **Cite appropriate references**

If you are using facts to support a case, state the source. Don't take someone else's ideas and use them as your own: you don't want someone pretending that your ideas are theirs – show them the same respect.

- **Spelling flames considered harmful**

Avoid the spelling flames. It starts out when someone posts a message correcting the spelling or grammar of a previous message. The immediate result seems to be that everyone on the CCS turns into a primary school English teacher, picking apart each other's postings for a few weeks. This is not productive; it tends to cause people who used to be friends to get angry with each other.

It is important to remember that we all make mistakes. There are also a number of people who suffer from dyslexia and who have difficulty noticing their spelling mistakes. If you feel that you must make a comment on the quality of a posting, do so by private e-mail, not a public posting. So be tolerant!

- **Manage the communication**

Good communication doesn't just happen. In any communication, you are doing two things: first, **exchanging information and feedback**, both facts and opinions. When communicating, be careful to separate out facts from opinions, and comments on the work in question from those which may relate to that person:

> 'I didn't like the changes that were made. I would have preferred you to have checked with me first.'

Depending on your perception of the other person, you may want to soften your feedback. This may add to the length of your message!

> 'I can see you have put in a lot of time and effort on these changes, and it is a shame we are not going to be able to accept them all. Next time, before committing yourself, check with me first and we can ensure your quality work is better focused.'

Secondly, you are **managing the exchange**. As we have already said, regular summarising: 'As I understand it, you were saying ...' or 'So, what we have agreed is ...', and the use of quotes from messages you are replying to:

> I told Fred that he should send me the
> specification before Friday, and he added
> he would include costs.

Good; Fred needs pinning down on dates.
Make sure you always get him to work to
one

helps the reader of your message make sense of the overall conversation and of what you are saying on this occasion. You can also:

— use SMART 'objectives' in communication – e.g. 'Will you give me your comments by Wednesday lunchtime?; or 'When you reply, make sure you give me cost and time details on both X and Y, but not Z'.
— agree a coding system for parts of the job – e.g. 'In Task 42, I wanted you to ... '.
— focus the communication (when needed) on the use of language – e.g. 'When I said "amalgamate" the two tasks, you understood "cut one task out". In future let's agree ... '.

● **Summary**

— Never forget that the person on the other side is human!
— Be careful what you say about others.
— Be brief.
— Your postings reflect upon you; be proud of them.
— Use descriptive titles.
— Think about your audience.
— Be careful with humour and sarcasm.
— Do not shout.
— Summarise what you are following up.

— Read all follow-ups and don't repeat what has already been said.
— Be careful about copyrights and licences.
— Cite appropriate references.
— Spelling flames considered harmful.
— Manage the communication.

Appendix K: Interview and appraisal questions

The questions here can be used both face to face and via e-mail.

OPEN

- To establish rapport (introductory questions or comments to establish relationship and put the remote worker at ease; social chit-chat).
- To explore broad background information ('Please tell me about ... ').
- To explore opinions/attitudes ('What do you think about ... ?'; 'How do you feel about ... ?').

PROBE

- To show interest/encouragement (non-verbal noises, 'Umm?', 'Er?', 'Oh?', plus supportive statements, 'I see ... ', 'And then ... ?', and repetition of some words to encourage further response). Note that these are mainly for face-to-face, but they can be effective in e-mail to reduce the formality of written text and convey the tone and feelings of the writer.
- To seek further information ('Why?', 'Why not?', 'How do your tasks now compare with those in your last project?', 'How do you mean?', 'Can you tell me more about it?', 'How would you feel if ... ?', etc.).

- To explore in detail particular opinions/attitudes ('Why do you feel that way?', 'You think that ...?', 'It seems to you that ...?', 'What do you think the other person was thinking?').
- To demonstrate understanding of and clarify the information already given ('As I understand it ...?', 'So what you are saying is ...?', 'What conclusions do you draw?').

CLOSED

- To establish specific facts/information ('Are you ...?', 'Do you ...?', 'How long did you take?', 'Who was involved?').

PLANNING

- To focus on the future ('So what will you do differently next time?', 'What are you going to have to work on to do better next time?').

(Developed from the idea of Ian Mackay.)

Appendix L: Record sheets

GROUP RECORD SHEET (SKILLS)	
ALL INDIVIDUALS, ALL SKILLS	
NAME:	SKILLS LISTED

INDIVIDUAL RECORD SHEET FOR EACH INDIVIDUAL
NAME: CV ATTACHED
SKILLS:
TASKS DONE:

Recommended reading

Anderson, N. and Shackleton, V. (1993), *Successful Selection Interviewing*, Oxford: Blackwell.

Corfield, R. (1993), *Successful Interview Skills*, London: Kogan Page.

Crabtree, S. (1991), *The Recruitment Workbook*, London: Kogan Page.

Date, M. and Iles, P. (1993), *Developing Management Skills: Techniques for Improving Learning and Performance*, London: Kogan Page.

Irwin, R. and Wolenik, R. (1986), *Winning Strategies for Managing People: A Task Orientated Guide*, London: Kogan Page.

Knell, A. (1989), *Employment Law and the Small Business: Business Enterprise Guide*, London: Kogan Page.

Lanz, E. (1991), *Employing and Managing People*, London: Pitman.

Ludlow, R. (1991), *The Essence of Successful Staff Selection*, New York: Prentice Hall.

Mackenzie Davey, D. (1989), *How to Be a Good Judge of Character: Methods of Assessing Ability and Personality*, London: Kogan Page.

Malone, M. (1993), *Discrimination Law: A Practical Guide*, 2nd ed., London: Kogan Page.

Nicholson, T. (1992), *Dear Boss: 52 Ways to Develop your Staff*, London: Mercury.

Northern Regional Management Centre (1994), *Helpcard 10*, Management Helpcards, Washington: NRMC.

Robinson, D. (1988), *Getting the Best out of People*, London: Kogan Page.

Sheal, P. (1992), *Staff Development Handbook,* London: Kogan Page.

4 Operational Issues

- Time management
- Administrative issues
- Meetings
- Equipment

Appendix

A. Guidelines for monitoring and controlling distribution of equipment

Recommended reading

TIME MANAGEMENT

Time management is about managing oneself and managing the demands placed on one.

Good time management is a necessity for everybody. We all organise our time, all the time (even if we do not do it consciously or very well), in our private and professional lives. We plan activities, we carry them out and we evaluate our actions. Unfortunately, often a lot of us get behind with work, miss deadlines or forget, and work piles up and we become frustrated.

In a traditional work setting, we usually work with others within the usual time parameters of nine to five, with coffee/lunch breaks. Planning of activities is often done for us – deadlines are continuously set, work is allocated and the watchful eye of colleagues and the boss are forever there. Motivation and time management are built in.

In remote working, effective time management has more to do with self-discipline and self-motivation. A highly organised remote worker gets on with his/her work, and works within a similar discipline to that of the office setting, though not necessary to the same patterns. The less-organised remote worker sometimes stumbles: work does not get done on time, paperwork piles up and letters go unanswered.

There are many aids to time management, from personal organisers and project planners through to the humble diary. Some are sold as systems that are highly structured and require great discipline to maintain; they are often expensive. While some people may find such systems workable, others do not. Indeed some may find they spend more time maintaining the system than doing the work!

In time management, it is important to find ways of managing oneself that are appropriate to one's personal style and needs.

ADMINISTRATIVE ISSUES

Every business requires some administrative support. When remote working operations form part of a larger organisation, then the administrative facilities and procedures already in place may be sufficient to cater for the needs both of remote workers and management. In certain circumstances, administrative tasks may be allocated to one or more individuals, working either from a central office or remotely.

Consider how each of the following functions will be performed.

RECEIVING INFORMATION AND GOODS

- How can the remote working organisation be contacted?
  - in person
  - by post
  - telephone
  - e-mail
  - fax.
- How will outsiders approach your organisation?
  - will customers/visitors come to see you? If so, do you need access to meeting-rooms (identify size and facilities required, bearing in mind the company image)?
  - do you need to provide meeting-rooms for remote workers (identify size and facilities required)?
  - where will the organisation's mail be sent from, and who will deal with it?
  - who will deal with phone calls, fax and e-mail messages

(is there a need for someone to be available during normal office hours)?

- How will supplies, equipment and stock be handled?
  - where will they be delivered to?
  - where can they be stored?

MAINTAINING RECORDS

- Information is an asset to any organisation, if it can be retrieved in a useful format:
  - what records need to be kept?
  - how will information be gathered?
  - how will information be stored, arranged and catalogued (i.e. what systems/media)?
  - what information needs processing or interpreting, and for what reasons (e.g. management information)?
  - what level of confidentiality is required?
- A remote working organisation will need to handle a range of information:
  - customer details
  - supplier information
  - marketing information
  - remote worker records (i.e. personal details, skills lists, performance appraisals, training records, etc.)
  - financial records (i.e. banking, insurance, budgeting, wages, salaries, invoices, orders, accounting records, etc.)
  - stock records
  - project data (i.e. specifications, plans, progress reports, etc.)

- work instructions and procedures
- legal information
- industry and general business information
- others.

ISSUING INFORMATION

- What information should be distributed and to whom?
  - how and what information should be circulated internally (i.e. to remote workers and management)?
  - what procedures will be used for communicating externally (e.g. to customers)?

Communication systems in remote working organisations merit careful consideration. Refer to Chapter 5 for more information.

GENERAL OFFICE FACILITIES

- Which of the following services will be required by the remote working management and workers?
  - handling mail/dealing with routine correspondence
  - taking messages
  - keeping diaries
  - secretarial support (e.g. word processing, desktop publishing, producing reports, taking minutes, etc.)
  - controlling stationery
  - reception of visitors
  - filing
  - photocopying/printing

- — ordering/purchasing
- — making travelling arrangements
- — organising events
- — others.

HANDLING FINANCE AND COMPANY ASSETS

- Will the remote working organisation have a separate financial function? If not, then consider:
  - — who will be responsible for the financial health of the organisation?
  - — how will the organisation's financial matters be dealt with (e.g. define procedures for banking, bookkeeping, making payments such as tax, insurance and wages, collecting payment from customers, budgeting, drawing up accounts and others)?
- What are the remote working organisation's assets and how will they be controlled?
  - — is there a need to maintain an asset register?
  - — what are the requirements for stock control and delivery?
  - — is there a need to issue, and control the use of, supplies and equipment (see Appendix A to this chapter for Guidelines)?
  - — who will oversee the maintenance of premises, equipment, etc.?

MEETINGS

THE PURPOSE OF MEETINGS

Meetings in any organisation take place in order to:

- exchange information
- consult others
- make plans
- review progress
- identify and solve problems
- take decisions
- negotiate
- brief people.

The requirement for remote workers to attend meetings will vary, depending on the nature of the work and the communication systems in use (see Chapter 5). By introducing guidelines for using particular forms of communication to deal with specific circumstances, it may be possible to reduce the frequency of face-to-face contact. In some cases, where a job involves simple, repetitive tasks, for instance, there may be no reason for remote workers to attend any meetings. However, where remote workers are organised into teams, meetings between members may be particularly important at the start of a job, in order to break down barriers and allow workers to get to know one another.

MANAGEMENT OF MEETINGS

In a remote working organisation meetings may need to be planned well in advance. The cost of attending meetings may account for a substantial proportion of the total cost of a job. If there is a need for regular meetings, then this must be taken into account during the scheduling of work, particularly when long travelling times are involved. The importance of managing meetings effectively may be considered greater under such circumstances.

The key points about meetings:

- purpose of the meeting stated clearly
- planning
- preparation
- time allocation
- follow the logical plan
- summary/agree actions
- follow-up/results
- make notes (in addition to notes taken individually, appoint someone to take minutes, circulate them and ask for feedback; this will ensure that important points are recorded and reduce the likelihood of misunderstandings).

EQUIPMENT

The requirement for different types of equipment will vary between remote working organisations. However, common to many will be the need for computer equipment, both for data processing and as a communications tool.

BUYING COMPUTER EQUIPMENT

Buy up-to-date equipment. The market moves very fast, and you should consider that most computers will need to be replaced or upgraded every two to three years to keep up with the demands of ever more sophisticated software and operating system enhancements. Higher-specification computers will be faster in operation. Speed is important where the work to be performed has a high graphics content, video work or is maths intensive. Ensure you are 'writing off' PCs and printers over two to three years in your accounts. This depreciation factor arises from the need to replace computers, and effectively spreads the cost in your accounts over the period.

TYPES OF COMPUTER

The two most common types of computer are IBM Compatible PCs and Apple Macintoshes. It is possible to share data and peripherals to some extent, although much equipment is specific to each type (e.g. keyboards, mice, expansion cards and some monitors). There is a trend for computers to be able to inter-operate to an ever increasing extent.

Apple Macs have always been popular for graphic design and publishing tasks. Many professional print bureaux are set up to handle output in Mac format, and may not be able to handle material from a Windows (PC) machine. If appropriate, you should check what your bureaux will accept and what 'format' they require.

WHAT ARE YOUR CUSTOMERS' REQUIREMENTS?

Your choice of computer hardware and software may be influenced by your customers' requirements:

- If you need to exchange data, consider what media can be used:

- printed material
- floppy or removable disk
- telephone (using a modem or an ISDN-equipped computer).

- Can you provide data in a format compatible with their hardware and software?
- There is a much larger base of software available for PCs than for Macs, and most business users will have Windows PCs.

HOW MUCH EQUIPMENT DO YOU NEED?

If you require a number of computers, it may be preferable to stick with one manufacturer, so that keyboards and mice can be swapped from machine to machine in the event of failure. However, as manufacturers change their specifications over time, this is really of limited help unless you are purchasing a number of computers at the same time.

PERIPHERALS

If peripheral equipment is to be shared (especially between Apple Macs and PCs), consideration should be given on how to maximise flexibility. This is also important if the equipment is likely to be moved from one location to another, with different uses on each occasion.

SCSI interfaces are especially useful here. Generally, six devices can be added to an SCSI-equipped PC (or a Mac). Basic PCs can often only support a limited number of peripherals (limited by expansion slots). SCSI allows all equipment, such as hard disk drives, scanners and back-up DAT drives, to be connected as the necessity arises and to be moved between machines easily. Separate software drivers will be required if the equipment is to be used on Macs and PCs, but this will be cheaper than buying duplicate equipment.

MODEMS

Modem communications can be problematic. Standards do exist, but manufacturers control their equipment in different ways. Unless you know a good deal about modems, pick well-known makes, with support and service facilities.

Don't buy anything slower than 28 800bps. Modem prices are falling, and if you do a lot of file transfer, get either 28 800 bps modems or consider ISDN digital telephone lines. ISDN is expensive at approx. £400 for installation, and you will need a terminal adaptor for your computer (normally in the form of an internal card) and suitable software, adding another £500 or so. However, if you will be sending large files regularly over the phone, it may be cost-effective.

Many modems now incorporate a fax capability for sending and receiving faxes directly to or from your computer; increasingly they offer voice processing too. This allows your computer to act as a fax machine and answerphone, as well as being used to transfer files or log in to other computer systems over the phone.

E-MAIL

If a number of remote workers need to communicate electronically, it may be worth setting up your own e-mail (electronic mail) system. There are a number of systems available at low cost.

PRINTERS

If you have PCs and Apple Macs, consider a printer that has both parallel interface and AppleTalk/and or Ethernet. There are many laser printers with both interfaces (HP, Epson, etc.).

NETWORKS

Will several people be working on the same project or wish to access shared data at the same time? If so, consider a network of computers using Ethernet.

MAINTENANCE

Take out a maintenance contract with PC purchases. It is generally cheapest to get on-site coverage when the purchase is made. The cost is easily recouped against the time and effort of packaging up equipment, plus carriage costs. Remember that the cost of the delay may be more significant than the cost of on-site warranty. Consider whether you need the equipment to deliver work to your customer. Your business may depend on it. Most on-site maintenance agreements cover any location, so even if your computers are in individuals homes, service can be provided. Check this before purchase, if necessary.

BACK-UP AND ARCHIVING PROCEDURES

Individual remote workers must be instructed to take regular copies of any work. On a temporary basis, floppy disks may be adequate; however, they tend to be mislaid or damaged by dust, etc. Use a more permanent method for important data, and to maintain records of where the data is held. A DAT (digital audio tape) is best for secure, long-term backup; if you have a network, this backup can be done automatically. DAT tapes should be stored away from computers after use to cover for the possibility of fire, theft, etc.

If you are just backing-up the data on the network, tapes are generally rotated such that several days' or weeks' data is not lost in the event of a failure of a disk drive, for example. If the data is to be archived, then consider having a separate procedure (and tapes!) for long-term storage. This frees up disk space and provides a better strategy for long-term storage of information that you may need to go back to in future months or years. Optical disks, or CDRs, are becoming increasingly popular for archiving work.

Keep a record of the location and names of files archived (or at least job numbers or customers' names), if handling large amounts of data.

KEEP AN ASSET REGISTER

If you own a number of machines, an asset register should list all equipment, with details of date of purchase, price, serial numbers, location, etc. You will need some of this information for your accounts; it is useful to mark equipment with unique numbers for the purposes of identification. This is especially true if equipment is to be loaned, or leased, to different people. You may also need to keep records of what is installed on each machine – it has been known for RAM chips to disappear!

Introduce a system requiring remote workers to sign for equipment and issue a receipt for each item as it is returned. See Appendix A for guidelines on monitoring and controlling equipment.

Appendix A: Guidelines for monitoring and controlling distribution of equipment

1. Update your asset register for each piece of equipment bought/sold and moved to a new location.
2. Mark each component of the equipment with an asset register number.
3. Prepare loan and return forms in duplicate, itemising each part of the equipment; for computers include internally installed cards and memory.
4. Check equipment on dispatch/return to ensure completeness (including cables, etc.) and proper operation.
5. Do a monthly inventory check.
6. Take note of manuals, software, disks, etc. issued.

If you are to hire or lease equipment, then refer to Chapter 8, p. 271, for a sample agreement format.

Recommended reading

Haynes, M.E. (1988), *Effective Meeting Skills,* Better Management Skills series, London: Kogan Page.

Peel, M. (1988), *How to Make Meetings Work*, London: Kogan Page.

Rae, L. (1983), *Meeting Management*, Maidenhead: McGraw-Hill.

Seekings, D. (1987), *How to Organise Effective Conferences and Meetings*, 3rd ed., London: Kogan Page.

Timm, P.R. (1988), *Successful Self-management: A Sound Approach to Self-management*, London: Kogan Page.

5 Communications Systems

INTRODUCTION

Central to the management of any business is **effective communication**. In Chapters 6 and 7 the more formal aspects of communication concerning work management are explained in some detail. Here we are concerned with the personal aspects of communication that are particularly relevant to remote working.

A key problem of remote working is the isolation of the worker from the management – this affects both the communication needs of work management and social contact. Just as the communication on work can be formalised, so certain practices can be introduced to encourage systematic communication to reduce the social problem of isolation.

In addition, in normal working environments interdepartmental and personal conflict can be managed by face-to-face communication. Other ways have to be found in remote working, and systematic communication is seen as an essential aid.

COMMUNICATION SYSTEMS

In addition to issuing and progressing work, the communication system exists in order to share information, and attitudes or feelings, in order to:

- gain or pass on knowledge
- get or give help
- teach and develop
- exchange and explore ideas
- persuade or negotiate.

Frequent and consistent communication is the key to:

- maintaining motivation levels
- reducing the feeling of isolation
- increasing feeling of attachment to the (main) company/office
- managing progress
- sorting out queries and conflicts.

The means by which communication in remote working can take place are:

- telephone
- fax
- in person
- letter
- via a personal computer and e-mail.

How to communicate should be left at the discretion of the persons involved, but it should be agreed before work starts. Different ways of communicating will suit different people in different circumstances. A telephone conversation may be welcome during an extended period of work at home; however, to inform on work progress or answer routine questions, fax or e-mail may be better.

Telephoning as a means of communication is by far the most common at present: it provides a chance for remote workers to convey their feelings, to complain, to ask for help and/or advice, and so on. It is not a good means for discussing particular aspects of a job in hand, unless both sides know exactly what they are talking about (technical particulars). A drawback of telephonic communication is, of course, the need for the physical presence of two people at either end of the telephone line (unless an answerphone is used). It should be remembered that too frequent phone calls can be disruptive to the work flow, and perhaps not everyone likes receiving phone calls from insomniac remote workers.

Faxing is a good means of sending documents or completed projects: it is quick and does not require the simultaneous physical presence of sender and receiver of the information.

Meetings are perhaps the best means of communicating with

remote workers, but tend to be the most difficult to arrange, requiring the physical presence of (at least) two people in the same place at the same time. Remote workers may have problems over travel or leaving dependants at home. Meetings should be arranged and negotiated by both parties.

Sending a letter is theoretically as effective as sending a fax; however, it is of course a much slower means of communication and not necessarily reliable.

Communication via a personal computer is fast becoming an everyday occurrence, with its easy access to the outside world. A great variety of information and advice is obtainable without the need for the physical presence of both sides during the communication, together with the ability to get answers or feedback almost instantly. It is, like fax and e-mail, the most effective **asynchronous communication** for remote workers, where **synchronous communication**, such as face-to-face or telephonic conversations, will be less used.

The frequency and form of communication will largely depend on the type of work that is being carried out, on the personalities of the individual remote worker and employer and on their respective needs.

In remote working, it is recommended that communication should ideally take place every day between employer and worker – preferably at an agreed time – and certainly not less frequently than once a week. Apart from helping work communication, this contact replicates to some degree the face-to-face contact found in conventional work.

A formal **reporting system** should be in place to aid supervision, progress monitoring and problem-solving. Because of the work setting, *ad hoc* communication with a remote worker is not always possible as it would be in the everyday setting of the office. The when and how of core communication channels should similarly be negotiated before work starts (for more detail see Chapter 3, p. 104). This does not mean, though, that informal communication is unimportant.

The employer must accept the fact that remote workers can contact them at almost any time.

The organisation needs also to remember to use other forms of communication which help workers to combat isolation, both social and professional:

- group meetings (project group meetings, annual get-togethers, residential days away, etc.)
- conferences
- regular circulation of briefing notes or newsletters, either by e-mail or in paper format (updates on work practices and new equipment are often compiled from remote workers' own experiences).

WORKER-TO-WORKER COMMUNICATION

In addition to communication channels between the employer and remote worker, and between remote worker and customer, we should not forget the need for communication between worker and worker.

In a regular work setting, such communications are often regulated in some way, for example, through interdepartmental meetings and committees, or are complemented through routine face-to-face contact in the corridor or canteen. With remote workers, the communications may rely more on the use of asynchronous channels, as already stated, when the choice of words has to carry all the additional meaning normally conveyed by body language and voice tone.

Remote workers need to know how to use e-mail facilities with people who will be relative strangers. The evidence suggests we have lost many of the letter-writing abilities of our parents, yet remote working often depends on them.

Also, while it is relatively easy for the employer to lay the ground rules for employer–worker communications, it is less so with worker and worker. A lot will depend on the workers wishing to communicate, and all also wishing to make the effort to manage the communication. This will be critical, for example, where two workers are each given responsibility for separate but interrelated pieces of work. They may need to negotiate and manage conflicting demands. In the regular workplace, the employer is more likely to pick up quickly on any problem and provide a brokerage role. With remote workers who cannot find ways to collaborate or cooperate, the problem may only surface when things become serious.

Management needs to ensure that in work involving two or more workers, careful consideration is given in inter-worker communication to:

- the when and how
- the identification of possible conflicts
- how conflicts can be resolved.

In addition to planned project or social meetings, the management can develop a culture of inter-worker cooperation through such means as a computer conferencing system (CCS) (Chapter 6 contains more about Teamworking, pp. 157–9).

COMPUTERS AND COMMUNICATIONS

Management should consider establishing a **computer conferencing system** as a core communication mechanism to link:

- remote worker and the company
- remote worker and remote worker
- remote worker and customer, if necessary.

The computer conferencing system can be accessed by the remote worker from his/her workplace using a PC and a modem. The remote worker dials into it, 'logging on', using the software provided. Management can stipulate that they expect remote workers to dial in every day; and the computer conferencing system can be used to:

- send and receive electronic mail
- read and post messages for public consumption
- read and send information concerning the organisation's projects.

A computer conferencing system is not designed to replace other forms of communication, particularly the use of telephone and face-to-face meetings. The value to the remote worker, however, is that the computer conferencing system is open 24 hours a day, so that they can access stored information and leave messages at times that are convenient. It only works, though, if they keep to a required discipline:

- dial in every day; get into the habit of checking messages routinely
- be an active participant and contribute as well as read; don't be passive
- be sensitive to other users.

Use of the computer conferencing system is an essential requirement of the company. At the beginning of a particular project, remote workers and management will agree:

- frequency of access to the system, and by whom
- information to be exchanged on every access
- how information on the project and worker activities will be stored.

The management should ensure either it or a designated person takes the role of facilitator at a **computer conference**: the role of the facilitator may combine the roles of chairperson, host, teacher and even entertainer!

The facilitator functions are listed below (after A. Feenberg, *Mindweave*):

- *contextualising* – setting norms and agenda
- *monitoring: recognition* (to assure participants that their contribution is valued and welcome) and *prompting* (to solicit comments from participants, either publicly or through private mail; might be formalised as 'assignments' in some conferences)
- *meta functions: metacommenting* (focusing attention on problems in the process of discussion: to remedy problems in context, norms, agenda, clarity, irrelevance and information overload) and *weaving* (summarising the state of discussion, to find unifying themes and points of disagreement in participants' comments – it encourages them and implicitly prompts them to pursue their ideas).

The computer conferencing system discussion can be used for a variety of types of communication that would normally be done face to face, namely:

- brainstorming (to generate new ideas)
- counselling and exchange of experiences and views; offering suggestions
- counselling services related to specific professions
- production/service (e.g. maintenance teams at a computer centre, user/customer support, etc.)
- project/development (e.g. research and development groups)
- communication channels for political groups, trade unions, etc.
- 'electronic college', with conferences devoted to teaching as well as administrative and social functions
- on-the-job training, with conferences devoted to teaching, administration and socialising, in addition to groups sharing personal and job-related experiences.

If the computer conferencing system is used effectively, it will implicitly:

- prevent unnecessary conflict and aid the management of conflict when it arises
- develop a stronger sense of 'groupness', even teamwork.

Careful use of language is essential in e-mail. The key principles are set out in Chapter 3, Appendix J, p. 105.

COMMUNICATION AND CONFLICT

Conflicts may arise not only because of problems of language, but also because people often do not know how to *manage conflict*.

A typical conflict scenario occurs as follows:

- Latent conflict with underlying issues
  which leads to
- trigger event or focus on the issue
  which leads to
- awareness of the conflict
  which leads to
- suppression or denial, *or*
- breakdown of the relationship, *or*
- manifestation of the conflict in the open
  which leads to
- conflict reduction.

Managers of remote workers need to be sensitive to latent conflicts and aim to remove them before a problem arises.

In simple terms, this can be achieved if the management and remote workers understand, trust and communicate openly with one another.

In remote working, the management will need to read between the lines of public e-mail discussions between workers and make a judgement as to whether an intervention is needed.

The conflict scenario, outlined above, suggests that if a conflict is triggered, it is better to bring it out into the open and resolve it, rather than suppress it. On a CCS, this can be done by the facilitator stopping all discussions, and beginning a separate one focusing on the conflict.

For example:

Manager: 'I want everyone to call a halt on their project tasks for a moment to work this problem through. Let me summarise things AS I SEE THEM'.

Susan: 'You said you wouldn't complete XYZ. Am I right in

thinking you don't want to do it because you are generally unhappy with the quality of Bill's work? Please specify the underlying problem, by 8 pm tonight.'

Bill: 'Your messages to Susan appear to me to be curt and offer little in the way of detail. Am I right in thinking that you have problems with the specification Susan gave you? Please specify the underlying problem, by 8 pm tonight.'

This simple example shows how the manager, using e-mail, can move a discussion of the management of the problem.

People in conflict need to understand a simple model:

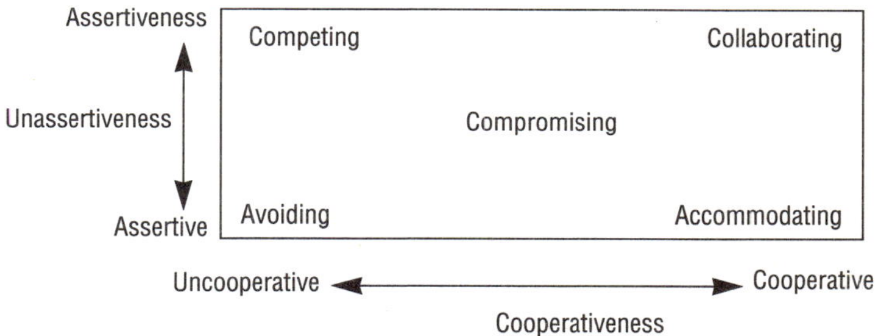

We talk about a 'win–win' situation, where both parties obtain what they want: this is a collaborative approach to conflict resolution and may require some imaginative thinking.

In some cases, one side or the other is obviously 'right' in the sense that what they want is likely to lead to a better outcome. Obviously, one has to consider for whom the outcome is better. The management is likely to have the business as a whole in mind. In such cases, people need either to learn to assert themselves without being aggressive or to recognise that their particular wants in certain circumstances are not particularly important – they need to learn to accommodate the other person.

Somewhere in between there may exist the need to

compromise, that is to give up certain wants. This may require some 'horse-trading' – a skill that perhaps few people are good at. Finally, some issues are just not that important and should be avoided.

In all of the above, the management of remote workers can play a key facilitation role, for example, running a brainstorming session on the CCS to look for win–win solutions. Or perhaps less important things can be distinguished from more important ones by adopting an objective, third-party viewpoint.

A structured discussion to resolve a problem might have a series of e-mail interchanges that follow these steps:

- define the situation – what is wrong?
- specify objectives – what needs to be put right?
- develop hypotheses – think of possible cause(s) of the problem
- get the facts
- analyse the facts – establish causes, do not concentrate on symptoms
- consider courses of action
- evaluate them
- decide and implement
- monitor implementation.

INFORMATION CAPTURE AND STORAGE

Because of the nature of remote working, much of the communication that takes place (i.e. over the telephone) may be lost. It is therefore imperative that all important information exchanged is recorded in a formal way.

An example of a formal means of communicating and reporting progress is the **progress report**, which ought to be completed regularly by the remote worker and submitted to the immediate boss as soon as practicable (e.g. daily use of e-mail). See Chapter 7, Appendix G, p. 232, for an example.

Records must be kept of all communications, appraisals and projects in an organised way (and one which is easy to retrieve). Using information technology makes this easier, particularly if some of the records are to be accessed and updated by the remote workers themselves. Ideally, remote workers *manage themselves*, and as part of this, manage the operational and information systems that relate to them.

The employer, however, needs to give some careful thought as to how central records will be both accessible by the employer and by the worker alike.

There are certain human resource records which help the running of the business, as well as allowing the organisation to shrink and grow according to market demand:

- *individual record sheets*, which are not dissimilar from personnel files.
- *group record sheets*, which list the individual skills of all remote workers employed (and on file).

For examples of these forms see Chapter 3, Appendix L, p. 114. These help assess the skills pool available, as well as the availability of appropriately skilled individuals to work on projects.

The management should consider extracting certain personal details from the various records to have 'pen-portraits' available to all relevant remote workers who are working on a project.

Recommended reading

Bland, M. and Jackson, P. (1990), *Effective Employee Communication*, London: Kogan Page.

Block, S.L. (1988), *How to Write a Staff Manual*, Better Management Skills series, London: Kogan Page.

Bone, D. (1988), *A Practical Guide to Effective Listening*, London: Kogan Page.

Davies, P. (1993), *Your Total Image: How to Communicate Success*, London: Piatkus.

Dening, P. (1988), *Readymade Business Letters*, London: Kogan Page.

Haynes, M.E. (1988), *Effective Meeting Skills*, Better Management Skills series, London: Kogan Page.

Northern Regional Management Centre (1994), *Helpcard 5: Communication*, Management Helpcards, Washington, Tyne & Wear: NRMC.

Peel, M. (1988), *How to Make Meetings Work*, London: Kogan Page.

Rae, L. (1993), *Meeting Management,* Maidenhead: McGraw-Hill.

Roberts, P. (1991), *Projecting your Image*, London: Mercury.

6 Managing Remote Working

- Supervising remote working
- Skill requirements for remote worker management
- Remote working organisation
- Teamwork
- Definition of roles
- Quality management
- Strategies, standards and guidelines
- Motivation
- Support

Appendices:

A. *Quality Assessment Rating Tool*

B. TNL's Software Quality – a policy statement

Recommended reading

SUPERVISING REMOTE WORKING

The focus of supervision in remote working is shifted from **process control** (activity) towards **results control**. In such circumstances, supervision is not as observable as that on-site. However, various control systems already in place within organisations could, and should, be used – for example, the existing project management principles, quality procedures (e.g. TQM, BS 5750), performance reviews, appraisal systems and reporting systems.

Close supervision (typical of on-site work) often emphasises and rewards activity rather than results (the employee looks busy).

Basic skills for remote managers include:

1. Planning work.
2. Delegating.
3. Setting timetables.
4. Assessing progress.
5. Giving performance feedback.

With on-site workers, these skills may get rusty as frequent contact with the employee often takes the place of more disciplined approaches to supervision.

1. *Planning work* – you must have a clear understanding of the work itself, and be able to figure out how long certain tasks will take to complete and how the pieces fit together. You need to know in advance what resources are required to complete a task (i.e. materials, files and information from co-workers/clients). The objective of this skill is to help do a good job of allocating work and make sure the staff are equipped to handle it.

With good planning, targets are SMART (*S*pecific, *M*easurable, *A*chievable, *R*elevant, *T*ime-based). See Chapter 7, on Project Management (p. 205), for more on planning work).

2. *Delegating* – this is a logical extension of the above. You need an effective way to hand over all assignments; it involves the following steps:

- break down the tasks into parts
- determine how well the worker can do each part and arrange for help or resources, if needed
- describe the task, its importance and the reason why the worker has been selected to do it
- discuss the amount of authority being delegated: where does the worker have discretion to act alone, and where should he/she seek approval?
- explain exactly what the desired end-product should be, in terms of due date, quality, quantity and any other relevant measures
- prepare and discuss a schedule of interim reviews.

In remote working, delegation is a form of contract between the manager and the worker; and it implies letting go and trusting the individual to complete the task.

3. *Setting timetables* – it is almost impossible to manage any kind of activity if the only contact/reviews are at the beginning and end of the project.

A timetable, in this context, is a list of deliverables, agreed and developed with the worker. It provides clear intermediate targets that help focus the work, and allows workers to make informed decisions about changes in work schedules and about their ability to take on more work. It necessitates visits to the main office, or from the manager to the remote worker.

4. *Assessing progress* – it is easy in the case of on-site work to track progress; in the case of remote working, it needs to be done either at fixed intervals or according to the key events, and to be linked to planned visits to or from the main office.
5. *Giving performance feedback* – this is especially important as remote workers do not have great opportunities for contact with the manager. Managers of remote workers must be aware of the need for feedback – as a motivating factor – to the worker. Since contact with remote workers follows a different path from that with on-site workers, it must be both detailed and two-way; it is required to:

- relate to the output of each job
- be limited to priority items: no one can fix ten things at once! (concentrate on top-priority items, move to others later; the aim is to try to make a person work better, not be perfect)
- be descriptive, not judgemental – tell people what you have observed, do not just label it using vague terms
- be ongoing – do not wait for an office visit, but pick up the phone, write a note or use electronic mail
- be timely – at the right time and in the right place.

See Chapter 3, on Human Resources (p. 57), for more on appraising workers.

It is strongly suggested here that remote working managers look at the MCI/NCVQ Occupational Standards for Managers as a useful framework of good management practice, strengthening those features specific to their remote working context.

TIPS FOR MANAGERS

- Spot trends, and look for the consistent trends in performance that clearly show above-average work: feed back and reward such work.
- Challenge the people – give 'stretch' assignments that call on the skills and potential of better workers.

Where the remote working operation is part of a larger organisation:

- Spread the word – keep higher levels of management aware of the high-quality work being done remotely.
- Assure exposure of the good performers to higher management – during periodic visits, if applicable.

Management of remote workers can be very time-consuming: the managerial span of control decreases by up to 50 per cent. More rigorous planning of work schedules, and setting measurable objectives, will help the process.

SKILL REQUIREMENTS FOR REMOTE WORKER MANAGEMENT

Managers of remote workers, in addition to the basic management skills set out by the National Council for Vocational Qualifications, need to pay particular attention to the following skills:

- Setting SMART targets (*S*pecific, *M*easurable, *A*chievable, *R*elevant, *T*ime-based):
    - briefing, reviewing, appraising

  - involving workers in long-term planning and short-term goal-setting
  - self-discipline
  - superior time-management skills.

- Quality control and supervision:

  - monitoring results, charting progress.

- Performing coaching and mentoring functions:

  - being on stand-by to provide extra training and encouragement.

- Helping with social contact.
- Negotiating contracts and an understanding of the legislation.
- Performing administrative tasks.
- Keeping communication channels open and functioning.
- Formalising procedures and simplifying them.
- Developing and maintaining trust and confidence in the worker.

  - motivating the worker
  - specific job motivation, as well as general motivation
  - offering a guiding hand and a shoulder to cry on.

- Computing, especially computer-based communications.
- Starting up a small business and dealing with tax and personal finance.

In choosing a remote working manager, the following factors need to be considered:

- the nature of projects – some have more slack than others
- managerial style – this will depend on the type of work being done and the skills of workers

- managerial skill
- willingness.

It is clear that managers of remote workers need to be *superior* managers. In some ways, they are running their own business with the requirement for all-round expertise.

REMOTE WORKING ORGANISATION

WHAT TYPE OF WORK CAN BE CARRIED OUT REMOTELY?

Some office jobs can easily be performed remotely; others can be broken down into parts, with suitable tasks allocated to remote workers. Such a breakdown of work requires a creative and flexible approach, together with careful planning, monitoring and control.

HOW CAN JOBS BE REDESIGNED WITH REMOTE WORKING IN MIND?

When dividing a job into tasks, it is useful to identify those that:

- are easily measured; particularly suitable for remote working are those with observable outputs, and in cases where jobs are difficult to measure, some acceptance criteria must be agreed between both sides before work begins
- require relatively little face-to-face contact
- are not dependent on frequent use of resources which cannot easily be made accessible to a remote site.

A remote working job profile should combine tasks, taking into account:

- the degree of professional expertise required
- the amount of time allocated to tasks
- the salaries appropriate to the tasks.

PLANNING REMOTE WORKING

The processes involved in setting objectives and assigning responsibilities are similar to those in any other organisation:

- identify what has to be done
- analyse requirements and divide the work into jobs
- break jobs down into tasks
- estimate and allocate time for each task
- select personnel to carry out tasks.

In a remote working organisation, the methods used to divide up work into jobs and tasks merit careful consideration.

WORK MANAGEMENT

When a job is broken down and allocated to a group of remote workers, management of the team effort requires a disciplined approach. Many of the techniques of conventional project management are useful in coordinating teams of remote workers.

TEAMWORK

HOW IS TEAMWORK RELEVANT TO THE ORGANISATION?

Remote workers may be organised into teams when:

- a job is too large for one person to carry out single-handedly
- a job consists of many tasks, demanding a range of skills
- a job does not require that individuals share tasks, but that they understand how their tasks contribute to the overall success of the job or project.

WHAT ARE THE CHARACTERISTICS OF A SUCCESSFUL TEAM?

Successful team members

- share a common purpose
- recognise that they can achieve more by working together than as individuals
- are confident in one another's abilities
- are happy to work for the good of the team, with the aim of achieving a common purpose
- combine efforts and share resources, including information
- understand their role within the team.

HOW CAN A GROUP OF INDIVIDUALS PERFORM WELL AS A TEAM WHEN THEY WORK REMOTELY?

The characteristics of a good team are the same in all organisations. Appropriate measures, however, must be taken to ensure that remote workers perform successfully as team members. Such measures include:

- Strong leadership, with a team leader who is a good manager, and who has experience of the problems associated with remote working and will ensure that workers understand that they are required to fulfil their roles as team members.
- A good communications system, which includes provisions for regular meetings between team members, with good electronic communications and procedures for reporting progress, problems, feedback, etc.
- A clearly defined quality policy, to ensure consistency of performance throughout the team.

ARE THERE ANY PARTICULAR PROBLEMS ASSOCIATED WITH TEAM WORKING IN THE ORGANISATION?

The following problems may present themselves:

- *Workers, particularly those who are self-employed, may be unwilling to cooperate with team members whom they see as potential competitors.*

  When building a team, this factor must be taken into account. It is imperative that team members are made aware of whom they are expected to work with. They should be clear at the outset that the work they undertake is the property of the remote working organisation, and that they are required to work towards the good of the team.
- *Team members may be wary of others they don't know, and because of limited contact, the time taken to build up team confidence may be prolonged.*

Experience of remote working team development indicates that in order to encourage individuals to become mutually supportive, they must get to know one another. To this end, team members should be encouraged to communicate; and they should be willing to allow information about their background and experience to be made available to other members.

- *Team members may be unable to perform their role successfully because of reliance on the performance of others.*
  Although this problem is not unique to remote working organisations, experience shows that when team members are self-employed, dependence on other workers can present a major stumbling-block. An organisation must issue clear guidelines on handling situations where team members are unable to carry out their duties due to the incompetence of others. The measures appropriate for dealing with such a situation may vary from implementing planned company procedures to instigating punitive action against the team member at fault.

DEFINITION OF ROLES

Within every successful organisation, members should be clear about their duties and responsibilities. In a conventional office set-up misunderstandings can often be spotted and dealt with early. In a remote working organisation, members are unable to learn about a company's normal working practices from observation. Self-employed workers, in particular, may be exposed to a number of different work systems or methods. It is essential that remote workers understand what is expected of them, and that mechanisms are in force to identify and resolve problems arising from misinterpretation of roles.

HOW CAN ROLES BE DEFINED?

The tasks relevant to a job in one organisation may differ widely from those associated with the same job in another. Simply labelling jobs with specific titles can be misleading. Roles should be clearly defined in terms of authority, responsibilities and the activities to be performed, as well as their overall significance within the organisation or team.

HOW CAN MISINTERPRETATION OF ROLES BE REDUCED?

Before starting a job, a worker should:

- be provided with a list of activities and responsibilities
- discuss the list with an appropriate representative of the remote working organisation and agree on where boundaries lie
- be clear that he/she has been given the necessary authority to take on the role
- be happy that he/she understands where the role fits into the scheme of things
- understand who to approach if he/she becomes unclear about what is expected.

HOW CAN MISINTERPRETATION OF ROLES BE DETECTED?

- By recording and reviewing how an individual spends his/her time.

- By recording problems and recognising when they arise from 'role-related' issues.
- By providing mechanisms for regular communication within teams of remote workers and between workers and management.

HOW SHOULD ROLES BE ALLOCATED?

An individual may be assigned to one or more roles at any time. If a remote worker is engaged to perform a job in isolation, then the definition and allocation of duties should indeed be straightforward. Where workers are members of a team, then it is necessary to take into account the balance of experience and personalities as well as skills, in order to arrive at a suitable allocation of roles.

HOW CAN PROBLEMS ASSOCIATED WITH MISINTERPRETATION OF ROLES BE RESOLVED?

Whatever systems of work are implemented within an organisation, procedures for recording and resolving problems are required to be clearly specified. Role-related issues should always be dealt with in accordance with appropriate company policy.

QUALITY MANAGEMENT

QUALITY SYSTEMS

Systems are developed within a remote working organisation to ensure that **quality policies** are defined and effectively implemented, to achieve a consistent and desirable quality.

DEVELOPING QUALITY SYSTEMS

A remote working organisation must identify the need for quality systems within specific areas of its operation and address that need appropriately. The development of quality systems is required to address issues relating to the organisation as a whole, as well as those concerned with the quality of work carried out by its remote workers. This can be achieved by:

- defining quality policies
- drawing up quality plans
- defining staff responsibilities with regard to implementing quality plans
- monitoring and reviewing the effectiveness of the quality systems.

QUALITY POLICIES

A quality policy needs to define the organisation's approach to quality, including its objectives and commitment to achieving consistent quality. Both staff and customers should be made aware of quality policies which affect them.

QUALITY PLANS

Plans are required to identify what needs to be done in order to implement the quality policies, such as:

- for each policy objective, define and quantify quality criteria
- identify the need for standards and guidelines (to specify how work should be carried out to the required level of quality; what tests are required to check the quality; and what should be done if quality levels are not achieved)
- draw up and issue standards and guidelines
- carry out procedures
- monitor and control the use of (and review the effectiveness of) standards and guidelines.

RESPONSIBILITY AND AUTHORITY

For each quality system, consider the following questions:

- who is responsible for managing the quality system?
- who has the authority to specify quality criteria?
- who is responsible for drawing up and issuing standards and guidelines?
- who takes responsibility for performing within agreed criteria?
- who has the authority to act in order to control performance?
- who is responsible for verification of the quality system?

VERIFICATION OF QUALITY SYSTEMS

Quality systems should be effective in areas of *real* importance to the organisation:

- Quality policies must be reviewed repeatedly, to ensure that they

- — refer to issues which concern the customer
- — refer to matters which make a significant contribution to the smooth running of the organisation
- — set realistic objectives.

- Quality plans should be reviewed, to ensure that they reflect company policies.
- Procedures need to be monitored, to ensure that quality plans are implemented.
- Records should be analysed to ensure that policy objectives are being met.

See Appendix A to this chapter for *Quality Assessment Rating Tool*.

SPECIFYING QUALITY CRITERIA

- At an organisational level, **Quality Guidelines** may take into account a requirement to meet standards set externally. For example, in some sectors BS 5750 accreditation automatically places an organisation in a favourable position, and some customers view it as a mandatory requirement. In other sectors, it may be more appropriate for a company to achieve alternative standards (such as the European Quality Award). Even where there is no requirement to meet such standards, an organisation may find it useful to base its own guidelines on the criteria laid out by external bodies.
- At a project level, **quality criteria** appropriate to the objectives should be defined and quantified. Consider what standards need to be achieved and how the level of achievement is to be measured or demonstrated. A **quality profile** may specify a range of acceptable values in the following categories:

- functionality
- speed
- capacity
- efficiency
- reliability
- maintainability
- safety
- adaptability
- cost
- time frame
- others.

DEVELOPING A REALISTIC APPROACH TO QUALITY OF WORK WITHIN THE ORGANISATION

In a remote working organisation, special attention needs to be given to the practicability of all aspects of a quality policy which affect its remote workers, with logistical considerations in mind:

- Introduce a quality policy with the emphasis on individual remote workers taking responsibility for the quality of their own work (see Appendix B to this chapter for example of TNL Software Quality Policy).
- Issue strategies, standards and guidelines to communicate quality requirements to remote workers (these must be made available before any contract is agreed, to ensure that workers understand what will be required of them).
- Include activities which contribute towards quality assurance in all work plans.
- Monitor the quality of work, where the most important issue for a remote working organisation is that monitoring

procedures are *effective* (they can be readily analysed to present a clear picture of what is happening), and *not too intrusive,* and also that they are *timely* (they must identify a problem as early as possible). Any mechanisms employed for monitoring purposes must be designed with the remote worker in mind:

(a) **Progress reports** should encourage reporting of any queries, problems (even if they have been solved) and opinions, as well as the total number of hours spent on tasks and lists of tasks completed. This type of reporting can help to identify recurring problems and highlight areas of uncertainty, and also demonstrate misunderstandings and record any useful suggestions which may be of benefit to other remote workers.

(b) **Regular inspections** should be carried out to ensure that progress reporting is accurate and that misunderstandings or errors are detected as early as possible. The method of inspection must be appropriate to the type of work performed, and to the location of the remote worker in relation to the inspector (e.g. large documents may be transferred electronically, local contractors may be engaged to perform inspections, etc.). Workers should be encouraged to accept regular inspection as an essential prerequisite of remote working.

(c) **Testing** should be carried out, as far as possible, by each remote worker in accordance with the philosophy of Total Quality Management. Workers must be encouraged to release work *only* when they are confident that it conforms to agreed quality criteria. There is no place in a remote working organisation for those who are happy to produce work that is *nearly* right, in the expectation that others will identify errors. This philosophy does not exclude the need for independent testing by others, such as supervisors and clients, but helps to rectify problems in the minimum possible time.

- Control the quality by:
  - identifying discrepancies between required quality criteria and the actual quality as early as possible.
  - identifying the reason for the discrepancy (e.g. use of faulty materials or machinery, failure unambiguously to define quality criteria, failure to issue adequate standards, failure to adhere to standards due to incompetence, misunderstandings, etc.).
  - determining the ideal outcome (e.g. the work is amended, repeated, etc.).
  - consider how the situation can be rectified and act accordingly (e.g. use an alternative supplier, review quality criteria/standards, train the worker, withhold payment, replace worker with someone else, etc.).

STRATEGIES, STANDARDS AND GUIDELINES

A **strategy** is a course of action which may be defined at any level – for instance, the organisational or project level.

In a remote working organisation, the development of strategies is useful in helping to assess the level of control required over various aspects of the business. A strategy document should describe the approach to be taken under pre-defined circumstances. An organisation whose workers are experts in their own field should be cautious about imposing unnecessary restrictions. Consider how each strategy contributes to the smooth running of the business or project.

Different types of strategy will be important to different types of business. Some may be particularly useful in communicating to workers the culture of the organisation itself (e.g. its approach to environmental issues); the list of strategies which could be drawn up is endless. Some examples are as follows:

- *Planning strategies* – outlining the processes to be followed, and issues to be addressed, when planning work (e.g. plan for additional meetings in the first phase of large projects to build up team confidence, as between remote workers, at an early stage).
- *Pricing strategies* – describing which pricing procedures should be applied to which types of work (e.g. some customers may be charged at a lower rate if they agree to make staged payments, linked to completion of parts of a job).
- *Feedback strategies* – specifying the need to provide different levels of feedback under certain circumstances (e.g. give feedback more often to remote workers who are new or inexperienced).
- *Payment strategies* – outlining preferred methods of payment for different categories of work (e.g. some remote workers may be paid a fixed rate on a monthly basis, followed by a bonus payment upon successful completion of a job, while others are paid only on delivery of a product).
- *Communications strategies* – describing appropriate methods of communications under a variety of circum-

stances (e.g. the need for formal procedures to control the distribution of mandatory standards to a team of remote workers).

Standards are specified procedures to be followed in order to achieve a successful outcome. They may be very detailed and are often mandatory; standards can be useful to a remote working organisation in the following ways:

- in circumstances where the quality of work must fall within strict boundaries
- to facilitate coordination and control of work
- to enable all workers to benefit from the knowledge of others within the organisation.

In some businesses where remote workers are experts in their field and who carry out tasks in isolation, there may be no need to impose standards, provided that other mechanisms are used to define and control the quality of their work.

In many cases, particularly where remote workers are also team members, the use of standards is vital to ensure that work achieves the required level of quality. No doubt some workers will perceive standards as imposing an undesirable constraint; however, experience shows that the need for formal control mechanisms within a remote working organisation is often greater than in a more conventional working environment. This is an important issue which should be discussed with every remote worker. The requirement to adhere to agreed standards should be included in workers' contracts.

Guidelines are procedures which help to achieve success. They give directions to workers on the normal way of performing a task. Guidelines are often issued in the form of checklists, itemising points which merit consideration under certain circumstances. They are a useful method of communicating ideas between an organisation and its remote workers.

MOTIVATION

Ideally, the remote worker is self-motivated. (What motivates them may be the deadlines to be adhered to, employment and income (or continuity of it), self-fulfilment, and so on.) Management can (or rather should) create a culture of motivation, to help remote workers and to help themselves. Afterwards comes the difficult task of *maintaining* motivation.

The main factors affecting motivation are:

- ability (intelligence, skills)
- perception of the job (what the individual wants to do or what he/she thinks is expected of them)
- influence (peer or family pressure and that of other groups the individual belongs to)
- the work itself (the extent to which it provides opportunities for achievement, the amounts of quality work and responsibilities).

Here are fifty tips for managing motivation, reproduced by permission of the author and the publishers from an article by Andrew Gibbons which first appeared in the December 1994 issue of *Training Officer*.

1. Find out what *really* motivates each of your people – you could always ask them!
2. Don't assume that money is the only (or even for many major) motivator.

3. Recognise that your own motivation will significantly influence that of others.
4. Praise often, but be direct and specific and, crucially, get in first – before the need is felt.
5. When giving praise, decide whether to do it individually or in front of others.
6. Remember the maxim: people will support what they helped to create – so involve others whenever possible.
7. Look for *de*-motivation, and take action without delay – work on causes not symptoms.
8. People are motivated to meet unsatisfied needs, so give more of what your team wants.
9. Don't assume that people are motivated by the same thing(s): we all want different things from life.
10. Motivators can change dramatically as needs are satisfied, so keep on top of these.
11. Delegate all you can, but make sure this is neither allocation nor abdication.
12. Remember you only *create* a situation in which people choose to be motivated.
13. All behaviour is motivated, so try to understand motives – don't just respond to behaviour.
14. Ensure all targets are SMART (*S*pecific, *M*easurable, *A*ttainable, *R*elevant, *T*imed).
15. Always give criticism constructively.
16. Assess each of your people's intention to stay against their commitment to achieve.
17. Learn to listen, and be different and better than the rest by showing active listening skills.
18. Seek ideas and suggestions – don't feel the need to know it all and make all the decisions.

19. Show trust and be open with your people – keep them in the picture, not in the dark.
20. Ensure everybody fully understands their roles and responsibilities.
21. Give advice and guidance where this is needed, and don't wait to be asked.
22. Be clear on what you expect of each person – agree and review key results areas.
23. Do all you can to make people feel good – that's really all it takes!
24. Give people the right to be *wrong* – this may mean a lot of tongue-biting on your part.
25. Discuss motivational issues with others and seek advice on how your skills can be developed.
26. Self-esteem (our need to feel good) is both precious and fragile, so handle with great care!
27. Give people variety and interest – feed back and exert some control over what they do and how.
28. Remember that supervisors and managers supervise and manage, leaders motivate!
29. Never take credit for someone else's work, indeed give credits to others where it may be yours.
30. Don't expect too much from people – despite your best efforts, some just won't want to play.
31. Be a better motivator than your boss – if necessary, manage how you want to be managed.
32. Wherever possible, try to allocate work on the basis of what is liked and done well.
33. Show real interest in people – this is a rare skill and will be appreciated: you may even get it back!
34. Be flexible in the closeness of your supervisor – know who needs to do their own thing.

35. Invite ideas and innovation, and use others' suggestions and ideas whenever you can.
36. Find time for people – this will be particularly appreciated when you've got your own troubles!
37. Recognise that the motivation to work will be influenced by external factors (e.g. domestic).
38. Understand the difference between *movement* (having to) and *motivation* (wanting to).
39. If a task is boring, make it bigger (by introducing variety, challenge and control).
40. If a task cannot be made bigger or more varied, ensure no one does it for too long.
41. Do all you can to *like* the people you work with, and show this.
42. Know the signs of *de*-motivation: absence, reduced productivity, lateness, and so on.
43. Don't feel threatened by developing your people and help them to get better.
44. Encourage teamwork, wherever possible, to build interdependence.
45. Watch out for tasks or role mismatch – a poor performer may just need a change of roles.
46. Take every opportunity to compare constructively your team against others.
47. Remember: the team that plays together stays together – so perhaps encourage social events?
48. Never tell lies, if you are unable to tell someone what they want to know – say why things are so.
49. Read all that you can on motivation, and try to understand why people are motivated.
50. Get the balance right between achieving task, team and individual needs (this is not easy).

In a remote working organisation where face-to-face contact is limited, the computer conferencing system is a useful way for workers to share their problems and frustrations. Management can initiate and facilitate discussions.

SUPPORT

Areas where remote workers may benefit from support and/or advice include the following:

- isolation (both professional and social)
- personal development and career development
- self-management
- self-employment issues, tax, health and safety, etc.
- knowledge of own role in the wider job/work context
- knowledge of own performance.

A number of strategies can be utilised here, namely:

- use off-site/on-site combinations of work, if possible
- consider using remote working flexibly over short periods of time rather than permanently
- provide access to e-mail to facilitate contact with others
- provide networks as focus for information exchange and opportunities to meet as a group
- provide training in self-management skills
- set up or use existing satellite offices, telecottages/neighbourhood work centres.

Appendix A: *Quality Assessment Rating Tool*

The *Quality Assessment Rating Tool* is a management tool to be used for evaluation of an organisation's approach to quality, including people issues. Give a score, say, in the range of 1 to 10, for each category. Use the results to highlight the priority areas for business development, and set targets for improvement. Repeat the assessment to monitor progress.

Quality criteria	**Evaluated score**
1. **Leadership**	
1a. Visible involvement in leading total quality.	
1b. A consistent total quality culture.	
1c. Timely recognition and appreciation of the efforts and successes of individuals and teams.	
1d. Support of total quality by provision of appropriate resources and assistance.	
1e. Involvement with customers and suppliers.	
1f. Active promotion of total quality outside the organisation.	

Average:

2. **Policy and strategy**
2a. How policy and strategy are based on the concept of total quality.
2b. How policy and strategy are formed on the basis of information that is relevant to total quality.
2c. How policy and strategy are the basis for business plans.
2d. How policy and strategy are communicated.

Quality criteria	**Evaluated score**
2e. How policy and strategy are regularly reviewed and improved.	
Average:	

3. **People management**
3a. How continuous improvement in people management is accomplished.
3b. How the skills and capabilities of people are preserved and developed through recruitment, training and career planning.
3c. How people and teams agree targets and continuously review performance.
3d. How the involvement of everyone in continuous improvement is promoted and people are empowered to take appropriate action.
3e. How effective top-down and bottom-up communication is achieved.

Average:

4. **Resources**
4a. Financial resources.
4b. Information resources.
4c. Material resources and fixed assets.
4d. Application of technology.

Average:

Quality criteria	**Evaluated score**
5. **Processes**	
5a. How processes critical to the success of the business are identified.	
5b. How the organisation systematically manages its processes.	
5c. How process performance measurements, along with all relevant feedback, are used to review processes and set targets for improvements.	
5d. How the organisation implements innovation and creativity in process improvement.	
5e. How the organisation implements process changes and evaluates the benefits.	

Average:

Total average:

Appendix B: TNL's Software Quality – a policy statement

TNL supports a quality management system based on the principles of Total Quality Management. The system requires that all those involved in a project accept responsibility for developing products which are *right first time*. The emphasis is on prevention, not detection.

In addition to satisfying the needs of the external customer, the quality requirements of internal customers (i.e. TNL management and other members of the project team) must be observed.

In order to achieve commonality of practice for management of projects and development of software products, members of the team are provided with guidelines and practical descriptions of TNL procedures. There will be instances when members will be expected to adopt mandatory standards. Undoubtedly, such standards may be seen by some as imposing undesirable constraints. However, it is our experience that a level of control is necessary to achieve the benefits of efficient, effective and economical software development.

The aim of TNL is to provide appropriate mechanisms, which will:

- Assist in the coordination and control of those activities necessary to achieve a desirable level of quality, at an acceptable cost.
- Enable procedures to be continually improved, to reflect experience gained and the changes in technology.

Recommended reading

Armstrong, M. (1986), *A Handbook of Management Techniques*, London: Kogan Page.

Armstrong, M. (1993), *How to Be a Better Manager*, London: Kogan Page.

Chapman, E.N. (1988), *The Fifty-minute Supervisor*, Better Management Skills series, London: Kogan Page.

Jenks, J. and Kelly, J. (1986), *Don't Do: Delegate*, London: Kogan Page.

Winckles, K. (1986), *The Practice of Successful Business Management*, London: Kogan Page.

7 Project Management

- The project management approach
- Risk
- Monitoring and controlling progress
- Configuration management
- Project planning
- Pricing a project

Appendices:

A. Project Initiation Report – Guidelines
B. Terms of Reference – Guidelines
C. Project Specification for Client – Guidelines
D. Project Specification for the Organisation – Guidelines
E. Risk assessment procedures
F. Risk Management Record
G. Individual Progress Report – Guidelines
H. Project review – Summary
I. Configuration reporting – Guidelines

THE PROJECT MANAGEMENT APPROACH

WHAT IS A PROJECT?

A project is any undertaking which has a finite life cycle. Its purpose is to establish and achieve objectives on time, within budget and to a level of quality which satisfies its users' requirements.

WHAT IS PROJECT MANAGEMENT?

Managing a project involves organising resources and implementing processes in order successfully to complete the project.

WHY IS PROJECT MANAGEMENT IMPORTANT TO A TELEWORKING ORGANISATION?

In many teleworking businesses, the majority of activities performed by remote workers are project based. Experience shows that even the most simple jobs benefit from a **project management approach**. Effective management of resources, such as people and facilities and systems, is essential to the successful completion of projects.

HOW DOES MANAGEMENT OF A REMOTE WORKING PROJECT DIFFER FROM THAT OF AN OFFICE-BASED PROJECT?

The responsibilities of a project manager in a remote working organisation are unlikely to differ significantly from those of the more conventional office-based organisation. However, the requirement for implementing *appropriate* methods of controlling progress, quality of work and communication can substantially alter the emphasis placed on some project management activities in the organisation.

WHAT IS A PROJECT LIFE CYCLE?

A project progresses through a number of distinct phases. Each phase falls into one of three broad categories:

1. *Pre-project activity* – defining and planning the project.
2. *Project implementation* – carrying out the plan.
3. *Post-project appraisal* – evaluating success and reviewing the process.

Pre-Project Activity – The Processes

Pre-project activity describes the processes carried out in order to arrive at a contract between the remote working organisation and the customer. These processes fall into the following categories:

- **Project initiation**
  - project conception
  - feasibility study.

- **Preparation of project proposal**
  - detailed specification of requirements
  - evaluation of solutions
  - production of project plan and schedule
  - project pricing
  - presentation of project proposal.
- **Project acceptance**
  - proposal agreement
  - proposal acceptance.

Project Implementation – The Processes

Project implementation includes all the processes involved in carrying out the project plan. Most activities will be specific to particular businesses; however, many fall into the following categories:

- Design of product(s) and/or service(s).
- Creation of product(s) and/or service(s).
- Project completion (installation, hand-over, etc.).

Post-Project Appraisal – The Processes

Following completion of a project, an appraisal is performed with the aim of evaluating the success of the project and reviewing the processes. The appraisal may involve:

- Success evaluation.
- Performance review.
- Follow-on activities.

WHAT ARE THE DUTIES OF A PROJECT MANAGER?

The duties of a project manager vary widely depending on the nature of the project and should be *clearly agreed at the outset.* However, a large number of management procedures can be applied to a wide range of projects. Many activities are common throughout the life of the project, while others are more clearly associated with one of its phases.

COMMON PROJECT MANAGEMENT ACTIVITIES

The following list identifies management activities which may be relevant at any phase. The significance of each one – and the methods appropriate – will depend on the nature and complexity of the project at different stages throughout its life cycle. For example, during pre-project phases for a small job, a manager might find the topic 'Tasks, roles and responsibilities' will be inappropriate; however, in a much larger project, 'Allocation of tasks, roles and responsibilities' during the same stage of the project cycle may be an important activity.

- *Plan this phase/stage of the project in detail* (refer to the section on Project Planning, p. 205).
- *Allocate tasks, roles and responsibilities* – project team members must be clear about what is expected of them and when.
- *Develop/review strategies then issue standards and guidelines* – relating to quality control, communication, team working, etc.
- *Provide motivation, support and guidance* – give encouragement to team members and advice on self-management, time management, administrative issues, etc.

- *Coordinate the team effort* – ensure that the work is being carried out efficiently.
- *Monitor and control progress, achievements and processes* – measure progress against plans, assess quality of work, evaluate effectiveness of communications mechanisms, appraise team and individual performance and identify problems.
- *Maintain a working relationship with customer(s)* – report on project status, involve customer in decisions and maintain customer commitment.
- *Provide feedback to the organisation and team members* – report on progress, explain decisions and advise as necessary.
- *Resolve problems* – resolve problems and conflicts, agree action plans.

PRE-PROJECT PHASES – PROJECT MANAGEMENT DUTIES

In this section, the processes are broken down into phases which identify a logical sequence of activities. The project manager may choose to perform the following activities, or allocate some of them to members of the project team. He/she should also consider the checklist of Common Project Management Activities, outlined above, and take action where appropriate. Whatever the circumstances, the project manager needs to be satisfied that the project objectives are clearly and accurately identified, and that they are attainable within the constraints specified.

1. Project Initiation

1a Project conception

A project is conceived when a decision is made to resolve a problem, or take advantage of an opportunity in an area of interest to the remote working organisation. It is important at this stage to keep a record of all discussions (see Appendix A for Project Initiation Report guidelines, p. 217). Consider the following:

- What are the essential requirements?
- Who are the interested parties (identify the decision-makers and others affected by the project) and what is their involvement likely to be?
- Where are the boundaries of the project?
- Who will pay and when?
- What is the budget limit?
- What is the timescale?
- Do we have access to members with the right sorts of skill?
- Are there any legal regulations to be considered?
- Is it likely that any capital investment is required (e.g. for expensive equipment)?

The answers to such questions will enable the remote working management to decide whether to reject the work – or take a closer look at the project. If the project is to continue, then it may be useful to define the terms of reference at this stage (see Appendix B for guidelines, p. 223).

1b. Feasibility study

Perform an initial investigation of the business requirements and evaluate possible solutions. Note that the importance of carrying out a detailed feasibility study will depend on:

- The nature of the work to be undertaken.

- Whether a similar project has already been successfully completed by the organisation.
- The relationship that exists between the organisation and the client. For example, in the most straightforward case when the project involves simple tasks, identical to those performed during a previous project for the same customer, then the feasibility study may be carried out informally and could be completed within a day. At the other extreme, for a large project, where there are many unknown factors at the outset, then a feasibility study may take several weeks to complete. In any case, the following checklist serves to highlight many of the important issues relevant at this stage:

  Identification of requirements
  — define the business objectives and technical aspects of the project.
  — what are the essential requirements and priorities?
  — where are the boundaries of the project?
  — what is likely to be the scale of the budget?
  — what is the timescale (are there any essential target dates)?
  — are there any legal regulations to be considered?

  Identification of users and/or business partners
  — who are the interested parties and what is their involvement likely to be?
  — who has the authority to make decisions and in which areas?
  — who will pay and how (are staged payments likely to be acceptable)?

  Identification of options
  — what are the key elements of the solution?

— can the solution satisfy the requirements?
— what would be the internal/external consequences of the solution (who or what other areas of work or systems would be affected)?
— what are the broad skill requirements?
— does the organisation have access to the right sort of workers; will they be available when required, or if not, how easy would it be to fill the shortfall?
— what other resources are required (equipment/services)?
— does the organisation have access to all the resources, and if not, how could they be acquired?
— what are the risks and how could they be managed (refer to the section on Risk, p. 196)?

Solution assessment

— compare options on their ability to satisfy essential objectives and determine which ones merit further consideration (for a large project it may be advantageous to think of the whole as a number of smaller projects, where only a proportion may be worth pursuing).
— draw up an outline project plan for options still under consideration (see the section on Project Planning, p. 205); are the timescales acceptable?
— estimate a price for each plan (refer to the section on Project Pricing, p. 212)
— select one (or more) preferred solution(s).

Feasibility assessment

— weigh up the advantages and disadvantages to the organisation in continuing with this project.

— what are the risks involved and are they acceptable?
— what does the organisation hope to gain from this project?
— will the project meet the expectations of all those involved?

Feasibility report

— record the findings of the feasibility study and the recommendations for further action. In some cases, the feasibility report may be of interest to the remote working organisation only as a basis for decision-making. In other instances, it may be presented to the client for the purposes of clarification of objectives and the agreement of terms for proceeding with the next stage of the project.

2. Preparation of a Project Proposal

The purpose of this stage is to define all aspects of the project as a basis for an agreement by all the parties concerned. Many of the activities will involve pursuing the issues considered during the feasibility study at a greater level of detail. The effort involved in preparing a project proposal will usually be greater for larger projects. As with the feasibility study, the time allocated to this stage will also depend on the nature of the project and the amount of previous experience from which comparisons can be drawn. (In some cases, the client may identify specific solutions to problems as prerequisites of the project, when an alternative approach might be more appropriate. In other cases, requirements may become invalid when a problem is handled differently.)

Clarify every point to avoid making incorrect assumptions and identify acceptable quality criteria. Consider the following:

- What are the essential requirements (i.e. identify, quantify and prioritise)?
- What other requirements are desirable (i.e. identify, quantify and prioritise)?
- Where are the boundaries of the project(i.e. as ideas develop ask whether the scope is too narrow and should one's outlook be widened)?
- What is the size of budget (and why)?
- What is the timescale (and why)?
- Are target dates critical?
- Are there any legal regulations to be considered (i.e. look at all the alternatives and consider the implications of failure to comply)?

2a. **Evaluation of solutions**

In view of the objectives and scope now identified, carry out a further investigation of feasible solutions. The degree of client involvement may vary at this stage. In some circumstances, clients are unable to appreciate the effects of their requests until the options are presented to them (prototyping tools used now may be invaluable, in reducing the need for later modifications):

Review of options

— what are the key elements of the solution?
— how can the solution satisfy each of the requirements within the boundaries of acceptable quality (i.e. rate the solution on its ability to meet essential functional and technical objectives)?
— what are the skill requirements?
— does the organisation have access to workers with the particular skills required?

— will workers be available when required, or if not, how easy will it be to fill the shortfall?
— what other resources are required (i.e. equipment/services)?
— does the remote worker have access to all the resources, and if not, how can they be acquired in time from a reliable source (i.e. consider availability, delivery, etc.)?

Final solution assessment
— evaluate the alternatives and make recommendations (as each stage of the evaluation progresses, some of the options may be discarded).
— compare the options on their ability to satisfy essential objectives and determine which merit further consideration.
— draw up a brief project plan for options still under consideration (see the section on Project Planning, p. 205); are the timescales acceptable?
— estimate a price for each plan (see the section on Project Pricing, p. 212).
— look at the options for including other non-essential, but desirable, requirements; consider the internal and external effects (e.g. on staff and on the client procedures, systems and organisation, etc.).
— select one preferred solution.

2b. **Production of project plan and schedule**
Having identified how the project is to be approached, it is now time to produce a detailed project plan and implementation schedule (refer to the section on Project Planning, p. 212, for more information).

2c. **Project pricing**

Project pricing involves calculating the cost of implementing the project plan and using this as a basis for fixing a price to be charged to the client (refer to the section on Project Pricing, p. 212).

2d. **Preparation of project proposal**

- *Project specification for client* – produce a clear and detailed description of the project objectives and how they can be achieved. The requirement of the specification is to ensure that all those involved share a common understanding of the issues and are clear about what is included and excluded from the project. Identify the **benefits** (see Appendix C for report guidelines, p. 225).
- *Project specification for the organisation* – produce an internal report which includes more detailed information about the work to be carried out (refer to Appendix D for report guidelines, p. 227).

3. **Project Acceptance**

The purpose of this stage is to discuss, refine and obtain authorisation for the project proposal and draw up contractual agreements with the client and remote workers.

3a **Proposal agreement**

- Discuss proposal with client.
- Evaluate or modify the proposal (if there are any fundamental changes to be made, then consider their likely effect on the project and repeat previous stages as appropriate).

3b. **Proposal acceptance**

- Negotiate terms of contracts with customers and remote workers.

- Approve contract on behalf of the organisation.

PROJECT IMPLEMENTATION – PROJECT MANAGEMENT DUTIES

Project implementation includes all the processes involved in carrying out the project plan and achieving a successful outcome. In order to execute the plan, a project manager will usually be required to perform most of the duties described as Common Project Management Activities. Additional activities carried out during implementation may include:

- *Manage the training of project team* – ensure team members receive adequate training.
- *Change control* – ensure that changes to the project proposal are evaluated, recorded, costed, authorised and implemented.
- *Configuration management* – ensure that mechanisms are in force for maintaining up-to-date information of different versions of a product or service.
- *Oversee provision of materials, supplies and services* – to meet the requirements of the project team and the customer(s).

Other management duties, at this stage, are likely to be specific to the business concerned but may come under the following headings:

- **Design of product(s) and/or service(s)** – ensure that the design conforms to appropriate specifications.
- **Creation of product(s) and/or service(s)** – ensure that the processes involved in creating the product/service are carried out efficiently.

- **Project completion**
  Deliver product/service – manage the installation/demonstration.
  Provide customer support – ensure that required level of support is received (e.g. training of customers' personnel, issue of supporting documentation/instructions).
  Identify and resolve problems – assess variances against customers' acceptance criteria, agree action plans and review objectives.
  Handover product/service to customer – sign off completed product/service.
  Provide backup – identify procedures for follow-on services, if applicable.

POST-PROJECT APPRAISAL – PROJECT MANAGEMENT DUTIES

During the run-down period which follows completion of a project, it is essential to evaluate the success of the project and review the processes. In addition to Common Project Management Activities, refer to the checklist below which identifies some important activities. A project manager will often perform these duties himself, in collaboration with senior members of the project team.

1. Success Evaluation

- *Evaluate project* – did the project meet the expectations of customers, the remote working organisation and the remote workers themselves on time, within budget and to a satisfactory level of quality?
- *Identify learning points* – identify strengths and

weaknesses, improve organisation standards and procedures and identify training requirements.

2. Performance Review

- *Appraise performance* – of individuals and team.
- *Recognise commitment and performance* – provide feedback/reward.

3. Follow-on Activities

- *Identify opportunities* – establish requirement for follow-on projects; identify marketing opportunities.
- *Maintain relationship with customer* – recognise the value of establishing a customer base; maintain contacts/customer information.

RISK

RISKS ASSOCIATED WITH REMOTE WORKING

All businesses face **risk**. Some are more vulnerable then others to particular types of risk. The remote working organisation is most likely to face risks commonly associated with its own type of business, rather than those which are attributable to worker remoteness. The actual difference between a remote working organisation and a more conventional business may lie in the methods used to control the risks.

RISK IDENTIFICATION

It is useful to identify sensitive areas of a business or project and define the types of risks that increase the likelihood of failure in these areas. Common types of risk are:

- accidents
- financial risks
- labour risks
- liability risks
- marketing risks
- natural disasters
- political risks
- sabotage
- social risks
- technical risks.

RISK ASSESSMENT

Having identified all significant risks, quantify them as far as possible. Consider the best, worst and most likely outcomes, in each case, and also the probability of their occurrence. Evaluate the risks by ranking them in order of priority – any with intolerable outcomes and high probability should have highest priority (see Appendices E and F for examples of Risk Assessment Procedures and Risk Management Records, pp. 229, 231). Determine what measures should be taken to deal with each risk, in accordance with company strategy.

RISK MANAGEMENT

All businesses should consider the benefits of developing

risk strategies. The same types of risk are likely to present themselves repeatedly and preferred methods for dealing with them can be devised. The following approaches may be considered:

- *Avoidance* – remove the source of risk. In cases where there is a high risk associated with an intolerable outcome, the only solution may be to turn away business.
- *Transfer* – shift the burden. When the risk is not so high, but the worst possible outcome is unbearable, transfer the risk to an insurer (or use a contractor to carry out parts of a job associated with an unacceptable risk).
- *Control* – accept a job which involves risks, but devise methods of dealing with them:
    - Reduce the probability of an undesirable outcome by detecting problems early and dealing with them. Achieve this by monitoring and planning remedial action.
    - Protect the business, by altering the sensitivity to a risk or limiting the damage (e.g. by introducing security procedures).
    - Devise contingency plans to minimise the effects of a risk once it has occurred. (In a remote working organisation, a procedure for effective communication of any contingency plans to *all* those involved must be implemented.)

MONITORING AND CONTROLLING PROGRESS

Work is monitored and controlled to ensure that it progresses according to plan, and that appropriate action is taken when necessary.

WHAT SHOULD BE MONITORED?

- Volume of work: are jobs being completed on time?
- Quality of work: is it meeting necessary standards?
- Changes: what changes have been made and how do they affect the job/project?
- Financial position: expenditure, income and cost of work compared with budget.
- Problems: how successfully are problems being handled and are there any recurring problems?
- Feasibility of continuing with the job/project: has something significantly reduced the likelihood of successfully completing the work?

METHODS OF MONITORING

To monitor any job it is necessary to collect information about what is happening and assess how it compares with the plan.

The following mechanisms merit careful consideration.

- *Progress reports* – these require to be completed by all remote workers. Those working as members of a team should be encouraged to report queries and problems (even if they have been solved) and opinions, as well as number of hours spent on tasks and lists of tasks completed. This type of reporting can help to identify recurring problems, highlight areas of uncertainty, demonstrate misunderstandings and record useful suggestions which may be of benefit to other remote workers. (See Appendix G for report guidelines, p. 232).

- *Meeting and demonstrations* – meetings cannot be replaced as a vital form of communication. Experience has repeatedly shown that, even in circumstances where all aspects of a job appear to be running according to plan, some misunderstandings are often detected *only* during meetings or demonstrations. The tendency to disregard the requirement for face-to-face contact in favour of other means of communication, due to logistical problems, is a mistake. Regular meetings should be included in all work plans.
- *Implementing procedures for other types of regular communication* – many aspects of a job can be monitored by less formal methods of communication. Telephone conversations or e-mail messages can often uncover problems which otherwise go undetected.
- *Inspections* – in order to verify progress reports and detect misunderstandings or errors as early as possible, regular inspections should be carried out. The method of inspection must be appropriate to the type of work performed and to the location of the remote worker in relation to the inspector (e.g. large documents may be transferred electronically; local contractors may be engaged to perform inspections; etc.). Workers should be encouraged to accept regular inspection as an essential prerequisite of remote working.
- *Testing* – all remote workers should carry out their own tests, in accordance with the philosophy of Total Quality Management. Workers must be encouraged to release work *only* when they are confident that it conforms to agreed quality criteria. There is no place in a remote work-

ing organisation for those happy to produce work that is *nearly* right, in the expectation that others will identify errors. This philosophy does not exclude the need for independent testing by supervisors and clients, but it does help to rectify problems in the minimum possible time. Procedures for testing the output from remote workers against predetermined criteria should be devised and rigorously implemented at the appropriate level.

- *Implementing approval procedures* – at key stages during the work process, it is often useful to obtain approval for a product or service, before investing further resources in the next stage of a job. 'Milestones' and approval authorities should be identified at the planning stage.
- *Change reports* – in some businesses, changes to a specification before the job is completed are the rule rather than the exception. Plans and specifications should be clear about the types and quantity of changes acceptable under the terms of contract. All remote workers should be required to report any changes to original specifications.
- *Problem reports* – all problems identified by remote workers should be recorded as soon as they occur. This ensures that potential delays are recognised and problems referred to the appropriate person. In cases where solutions are found immediately, the reports can help to identify recurring problems and record methods of handling situations which may be of benefit to other remote workers.
- *Financial reports* – in order to control budgets, regular reports of expenditure must be made. They need to be precise and easily understandable. They should also include direct comparisons with the planned budget and identify the cause of any deviation.

- *Audits* – these are independent appraisals of the status of a job and the prospects of its successful completion. They can be of great benefit to a remote working organisation, particularly when projects involve a number of remote workers organised as a team. They can also provide useful feedback to management on the success of strategies, standards and guidelines.
- *Progress reviews* – involve key members of the project team, or organisation, and should be performed regularly as a means of evaluating the information collected. (Refer to Appendix H for Project Review summary, p. 236).

HOW CAN THE WORK BE CONTROLLED?

To control a job, you should:

- **Define strategies, standards and guidelines** (see Chapter 6, p. 167, for more information).
- **Act on the information gathered via the monitoring process**
  - Identify discrepancies between planned and actual progress.
  - Evaluate the cause (e.g. failure of workers to conform to standards, machine failure, accident, etc.).
  - Determine the ideal outcome (e.g. workers conform to standards; machine failure reduced to 10 per cent; risk of accident removed; etc.).
  - Consider alternatives and act (e.g. send worker on training course; replace machine; nail down loose carpet; etc.).

- **Review/update the work plan**
  - Where has the discrepancy between the planned and actual progress occurred?
  - What action is necessary or possible and when (e.g. a worker can be trained and resubmit the work within three days, or a new machine can be ordered and delivered within three weeks)?
  - Produce a list of tasks to be completed.
  - Determine how the schedule is affected (e.g. shortfall can be made up during a period of 'slack').

CONFIGURATION MANAGEMENT

WHEN IS CONFIGURATION MANAGEMENT IMPORTANT IN A REMOTE WORKING ORGANISATION?

Configuration management is the process of identifying and controlling versions. It is useful for

- keeping track of actions and events which result in change.
- identifying the status of items such as products, components, documents, etc.

It is essential in a remote working organisation when items are simultaneously updated by more than one worker.

CONFIGURATION RECORDS

Configuration management necessitates that each item has a unique identifier and that a library of items and versions is maintained. The information recorded for every item must be sufficient to allow for each version to be traced to its source.

The content of each record will vary, depending on the item in question, but may include such things as:

- date
- produced by
- verified by
- specification identification and version
- associated products and versions
- associated documents and versions
- tools/machinery used and versions.

Refer to Appendix I for examples of configuration records for software development, p. 237.

CONFIGURATION MANAGEMENT PLANS

Plans should specify activities, allocate responsibilities and define procedures. Activities identified should be incorporated into work plans and schedules (see the section on Project Planning, p. 205).

CONFIGURATION MANAGEMENT PROCEDURES

When drawing up procedures, consider the following:

- Which items should be covered by configuration management?
- When (at what stage) does an item come under configuration control?
- What tools or techniques should be applied?

- Who is responsible for configuration reporting?
- Who has access to configuration records?
- What type of configuration reports are required, by whom and why?

PROJECT PLANNING

PLANNING IN PHASES

A project should be specified in terms of its objectives and the strategy for achieving them. For planning purposes, a project may be divided into phases, so that a phase is completed when a goal is achieved or a 'milestone' reached. This facilitates the estimating and monitoring of progress. Note that objectives should include financial, management and quality targets, as well as user requirements).

JOB IDENTIFICATION – WHAT NEEDS TO BE DONE?

For each phase of the project list targets (or milestones) and identify all the jobs which need to be carried out. These may include activities specific to one phase of a project, as well as those which are common to more than one phase.

JOB BREAKDOWN

In many cases, it is useful to break down jobs into tasks. The level to which the breakdown is required will depend on the nature of the project and degree of expertise both of the planners and remote workers. For example, on a project where large amounts of work are to be assigned to experts, then a whole job may be allocated to one remote worker. In such cases, the project leader may require that the worker produces his/her own task breakdown and report on progress with reference to those tasks. On smaller projects a task may be a small, measurable activity allocated to a team member.

DEFINING TASKS/ACTIVITIES

See Appendix J for a sample Task Sheet (p. 241), and Appendix K for an example of a Job Breakdown, p. 242).

In order to achieve a high degree of accuracy in planning, monitoring and estimating, it is often useful to identify tasks which:

- Require no more than five man days' effort.
- Have a beginning and end – try asking:
  - — how can it be measured?
  - — does it depend on completion of other tasks?
  - — are there other prerequisites?
  - — what are the completion criteria?
- Can be assigned to one team member.
- Can be defined specifically – try asking:
  - — what is the task?
  - — can it be quantified?
  - — what does it include/exclude?
  - — what preparation is required?
  - — what level of quality is required?
  - — what documentation is required?

— what support/resources are required?
— who will provide or pay for the support/resources?

CATEGORIES OF TASKS/ACTIVITIES

Jobs may be broken down into tasks which fall into the following categories:

1. Preparatory work.
2. Productive work.
3. Quality assurance.
4. Administration and support activities.

CHECKLIST OF TASKS/ACTIVITIES

It is often easy to identify the major elements of productive work associated with a particular job. However, *other tasks* (which are vital to the successful completion of the job) must not be forgotten. It may be useful to check that the following activities are taken into account for *every* remote worker.

1. *Preparatory work*

- Estimating, planning and scheduling own workload.
- Training and research into products/services (e.g. learning how to use new equipment, tools, etc.).

2. *Productive work*

- Repeating/amending work. Depends on the nature of the work, but can often be reduced by careful preparation, planning, monitoring and control; although the likelihood of changes being made to the job specification should be considered and a contingency element introduced, if appropriate.

3. *Quality assurance*

- Completing progress reports. If workers are required to evaluate their progress and report on problems and queries, this may take longer than expected.
- Attending meetings (re briefing, planning, review, feedback, problem-solving, etc.); in a remote working organisation, it is advisable to plan for regular meetings. Take into account preparation and follow-up, as well as travelling time when planning for meetings.
- Testing of own work (planning tests, carrying them out and recording the results).
- Demonstrating/prototyping (include planning, preparation and evaluation of feedback, as well as carrying out the demonstration).
- Dealing with problems or changes (identifying and evaluating solutions, etc.).
- Documenting and record-keeping (maintaining information about how a job was done and details of versions, etc.).
- Appraising performance and procedures.

4. *Administration and support activities*

- Communication (e.g. to coordinate effort and discuss key issues etc.; a few telephone calls could take up half a day each week).
- Setting up equipment.
- Purchasing supplies.
- Photocopying, DTP, etc.

Note that some remote workers may have no easy access to office supplies and facilities – it could take someone half a day to photocopy a few diagrams.

ALLOCATION OF TASKS AND ESTIMATING AND SCHEDULING

Allocation of work, estimating the time required to complete jobs and scheduling are dependent on one another. A remote worker who may appear to be most suited to a job or task may be unavailable for work at the time required, whereas another worker, who takes longer to complete the same work, may be allocated to the project because of availability.

Allocation of Tasks

Draw up a task list and consider the following:

- what skills are required to perform the jobs/tasks?
- who is available and when?
- what is the most effective use of an individual's time, taking into account their skills and the rate they charge?

Estimating Time To Complete Jobs

(See Appendix L for sample summary of Project Plans, p. 243.)

For each job or task:

- estimate the effort (man-days) required
- calculate the elapsed time (based on the job/task specification, the team member allocated and the number of hours per day/week available).

In a remote working organisation, one or more workers may be asked to provide an estimate of the effort (and cost) involved in completing jobs or tasks; this has several advantages:

- It may help to clarify what a job entails (if two workers present widely differing estimates, then the specification of the job/task may be too vague).

- It can reduce the risk to the organisation, in cases where a remote worker agrees to a timescale and cost for completing a job/task.
- It can contribute, in cases where the remote worker is experienced, to the accuracy of the overall project estimate.

Scheduling

The following steps should be followed in order to produce a schedule:

- *Establish precedences* – use a diagram to illustrate the sequence of tasks, the dependence of tasks on each other and the elapsed time required to complete the tasks. A number of methods and tools are available, including Gantt charts, which are particularly useful for planning events which follow in sequence and in projects where several activities are carried out simultaneously.
- *Produce a project schedule* – provide a schedule of tasks taking into account the following:
    - proposed start date
    - availability of suitable workers
    - delivery/availability dates associated with external factors (e.g. due to interaction with other projects or restrictions on the availability of equipment or facilities)
    - target completion date
    - warranty/support period (following handover to the customer)
    - holidays
    - seasonal factors
    - any industry variations in standard working practices.

A number of alternative schedules may be produced, in cases where several options exist for allocating or sequencing tasks. The costs of each alternative should be calculated (see the section on Project Pricing, p. 212). The options are assessed and a balance achieved in order to make most effective use of resources within the time and budget available.

Summaries of the schedule can be drawn up in the form of Milestone Plans (see Appendix M, p. 259).

USING COMPUTERS FOR PLANNING AND SCHEDULING

A variety of **project management software** is available off the shelf, and much of it concentrates on planning, scheduling and assessing progress.

The advantages of using different types of computer tool will depend on the size of the project (in terms of number of activities, timescale and resources) and the nature of the work. For example, in a small project involving two team members working independently on a series of simple tasks the most suitable and cost-effective method of planning and scheduling is likely to be a manual one. However, in a project involving ten team members, whose activities are linked, any small changes to a plan may result in many hours spent adjusting actions. In such cases, the following advantages of using computer software may be evident:

- data can be processed quickly and accurately
- planning, scheduling and monitoring processes can be easily repeated to reflect changes
- management information can be readily produced.

See Appendix N, p. 261, for guidance on evaluating computer software.

PRICING A PROJECT

WHEN SHOULD THE PRICE FOR A PROJECT BE CALCULATED?

Price estimates can be produced at various stages of a project. It is not always advisable to wait until a detailed project specification is produced before making an attempt to price a job. For example, a broad figure produced at the start of a feasibility study may be useful to give an indication of the size of the project, and it can be a vital factor in deciding whether or not to continue the project before incurring unnecessary expenditure.

WHAT PROCESSES ARE INVOLVED IN PRICING A PROJECT?

- Calculate the cost of implementing the project plan.
- Determine the marketing strategy (and associated costs).
- Determine what proportion of the costs will be passed on to each customer.
- Apply a profit margin to arrive at a final price.
- Identify subsidiary costs (the customer may incur other costs which are not included in the price charged by the remote working organisation).

THE COSTS OF IMPLEMENTING THE PROJECT PLAN

1. **How can accurate estimates of cost be calculated?** Use the best available information. Whatever level of detail is accessible, it should *all* be taken into account, to make the estimate as accurate as possible. The temptation when pricing a project in its early days is to take an educated guess at the cost of all aspects of the job (including those for which detailed information

is readily available). It is advisable to recalculate estimates as a project progresses. The accuracy of the estimate can be improved as more detailed information is uncovered.

The best estimates are produced when they are based on accurate information taken from records of similar projects:

- Keep detailed records of project costs, review them after every project and use this information as a basis for future estimates.
- Look at industry figures for guidance (where an organisation has limited experience of a similar project, valuable information may be gleaned from industrial reports).
- Identify where the project under consideration differs from previous projects. Quantify the differences and evaluate the variation in costs.

Identify the risks associated with any estimate, and the reliability of the cost of each item. Determine whether any action can be taken to reduce the risk. If a cost is subject to variation, produce estimates for the 'best', 'worst' and 'most likely' outcome.

2. **What are the components?**

Labour – use the estimates of effort required to complete each job (taken from figures identified during the project planning process) – together with remote workers' rates of pay, to calculate the total labour costs for the project, and/or invite remote workers to submit proposals and estimates for jobs.

Overheads (e.g. National Insurance, holiday pay, etc) – if remote workers are employees.

Expenses – for each job or phase of the project, identify and quantify. (Note that you should have a clear understanding of exactly which items are included or excluded in the remote workers' rates of pay; e.g. workers may be responsible for their own local travelling and telephone calls, but business trips may be charged separately.)

Equipment and materials (to be supplied to the customer) – for example, computer hardware, electrical equipment, paint, nails, etc., plus delivery and installation costs, if not included elsewhere.

Supplies and maintenance (equipment, tools, office supplies, etc. required to carry out the project work, if not provided by individual remote workers) – in cases where these can be re-used, a proportion of the total cost may be included here.

Equipment hire (if not included in remote workers' rates of pay).

Project management – usually calculated as a percentage of the total project cost.

Training of remote workers (including induction).

Insurance.

Administration (including support services and other hidden costs; e.g. hire of meeting rooms) – if the organisation is able to provide these facilities, it may impose a charge (e.g. a fixed amount per week) for the duration of the project. Otherwise each item should be identified and costed at commercial rates.

Contingency – if appropriate (see below).

3. **Is there a need for contingency?** It is often said that contingency does not improve the value of an estimate, but demonstrates the lack of a systematic approach. In some businesses, where projects follow a well-established pattern with containable risks, this may be true. However, in many situations costs are known to be calculated on the best available information (at the time) and it is recognised that there is an element of uncertainty involved. It is also well established that the majority of people

demonstrate a tendency to underestimate. One method for dealing with such circumstances may be to quantify the degree of uncertainty and introduce a contingency element as a security measure.

DETERMINE THE MARKETING STRATEGY AND ASSOCIATED COSTS

For the purposes of project pricing, the following issues should be resolved:

- How many customers will there be?
- How will the product/service be marketed?
- What will be the costs associated with sales and marketing?

Determine what proportion of the total costs will be passed on to each customer

If there is only one customer, then the total project costs should be included in the final price.

If there is more than one potential customer, then determine how the costs of the project (including sales and marketing costs) will be apportioned.

Apply a profit margin to arrive at a final price

Consider the following:

- Is it company policy to apply a fixed percentage profit?
- What will the customer pay (i.e. you may lose the job if you are not competitive)?
- Are there any other benefits to the organisation in getting this job, and which may affect the pricing policy (e.g. gaining a foothold in a new market)?

Identify subsidiary costs

A customer may incur other costs which are not included in the project price, but which may affect his decision to proceed. These should be included if you are required to provide a cost–benefit analysis; they may include:

- Cost of purchasing or leasing additional equipment (e.g. computer hardware/ software, etc.).
- Expenditure associated with installation (e.g. room alterations, furniture, provision of services, etc.).
- Operating costs (e.g. power, stationery, system operators and managers, etc.).
- Maintenance costs.
- Cost of time lost due to training and introduction of new facilities or procedures (including a handover period when new and old systems may be used simultaneously).
- Cost of preparatory work (e.g. initial loading of data on to a new computer system).
- Cost of time invested by the customer in specifying requirements, providing information, acceptance testing, etc.
- Cost of evaluation of product/service (e.g. performance monitoring).
- Costs arising from changes in customers' requirements.

Appendix A: Project Initiation Report – Guidelines

PURPOSE OF THE PROJECT INITIATION REPORT

This report provides a mechanism for recording information about a client company, personnel within the company and contact made, with the prospect of initiating a project.

COPIES

This report is for internal use only. Copies should be made available to management, the project manager and other members of the project development team (at the discretion of the project manager).

SECTIONS INCLUDED IN THE COMPLETED REPORT

(a) Project name – if a name has been assigned.
(b) Project identifier.
(c) Client company name – full name of potential client company.
(d) Phone/fax/e-mail of relevant client or office.
(e) Address – full address of relevant client/office (including postcode).
(f) Raised date/by – when report was raised and name of manager who initiated report.
(g) Introduced by – name of person supplying 'lead'

(commission may be offered to persons introducing new clients).

(h) Reason – why was initial contact made?

(i) Feasibility study date – when initial contact leads on to a feasibility study.

(j) Contact date/file reference – date contact made, and file reference of further details of contact with client, if relevant.

(k) Name of representative of client company.

(l) Position of representative within the client company.

(m) Phone/ext/fax/e-mail of contact, if different from company details.

(n) Summary/other – brief details of discussion or contact; other relevant information (when and where available, meeting times, etc.).

Project Initiation Report – header page

(For internal use only)

Project Name	
Project Identifier	
Client Company Name	
Phone/Fax	
Address	

Raised date/by	
Introduced by	
Reason	
Feasibility Study date	

Contact date/File ref.	
Name	
Position	
Phone/Ext/Fax	

Summary/Other	

Project Initiation Report – continuation page

Contact date/File ref.	
Name	
Position	
Phone/Ext/Fax	
Summary/Other	
Contact date/File ref.	
Name	
Position	
Phone/Ext/Fax	

Summary/Other	
Contact date/File ref.	
Name	
Position	
Phone/Ext/Fax	
Summary/Other	

Appendix B: Terms of Reference – Guidelines

PURPOSE OF TERMS OF REFERENCE DOCUMENT

To define the key elements of a job or project as a basis for agreement on its scope and objectives. It aims to reduce misunderstandings between interested parties before progressing to the next stage.

COPIES

Copies are to be circulated internally and externally to all interested parties (including clients, project manager and organising management).

ITEMS TO BE INCLUDED IN THE COMPLETED REPORT

(a) Project name – if a name has been assigned.
(b) Project identifier.
(c) Date – when document was produced.
(d) Client company name(s) – the name(s) of potential client company.
(e) Context – background to the project.
(f) Essential requirements – a list of all the major requirements (with references to the originator(s) where relevant).
(g) Other requirements – list of desirable, but non-essential, requirements (with references to the originator(s) where relevant).

(h) Scope – where the boundaries lie and why.

(i) Constraints (e.g. costs, timescales and use of specified resources).

(j) Interested parties – list of anyone involved (in addition to clients, this might include potential partners, consultants, bodies from whom funding may be available, etc.).

(k) The next step – explain what will happen once the terms of reference have been agreed (e.g. 'A detailed feasibility study will be undertaken by ... This will involve a more detailed study of requirements and consideration of alternative solutions. Cooperation and input will be required from ... A feasibility report will be produced dd/mm/yy').

(l) Authorised by/date – all those involved will be required to give their approval to this document before the next step can begin.

Appendix C: Project Specification for Client – Guidelines

PURPOSE OF THE PROJECT SPECIFICATION FOR CLIENT

This report records information gathered during the proposal preparation stage of a project development. It defines all aspects of the project of interest to the client as a basis for drawing up contractual agreements.

COPIES

This report is intended for internal and external use. It provides a mechanism for recording objectives and how they will be achieved, and it enables the client and remote working organisation to reach a mutual understanding of the major issues involved in the project.

SECTIONS TO BE INCLUDED IN THE COMPLETED REPORT

(a) Contents.

(b) Summary – the purpose of the project, what it will achieve and how, plus a summary of the benefits.

(c) Detailed specification of requirements – specific and unambiguous information covering all the requirements identified: a description of how the objectives will be achieved; what the inputs and outputs will be; what the client will need to do and what he will get out of it; and the criteria for client's acceptance (include samples of products, equipment list, etc.).

(d) Project schedule – a breakdown of the project plan in terms of 'milestones' of interest to the client, with reference to critical target dates.

(e) Project price – a breakdown of the price that the client will pay, with any conditions attached, e.g. staged payments (include here a description of other costs to the client not included in the price).

(f) Project justification – an analysis of how the client will benefit from the project.

Appendix D: Project Specification for the Organisation – Guidelines

PURPOSE OF THE PROJECT SPECIFICATION FOR THE ORGANISATION

This report should be prepared in conjunction with the project specification for the client. It records all additional information gathered during the proposal preparation stage, which is not included in the client's specification. Together with the specification for client, it provides a mechanism of recording the information gathered, the analysis of issues and decisions made.

COPIES

This report is intended for internal use. It contains vital information to be used as a basis for the next stage of the project. It also provides information required for the purpose of reviewing the project and improving standards and procedures.

SECTIONS TO BE INCLUDED IN THE COMPLETED REPORT

(a) Contents.
(b) Summary.
(c) Detailed specification of requirements – any additional information regarding requirements, such as:

- which user requested what

— priority lists
— why some requests will not be satisfied
— reference to client documents, etc.

(d) Evaluation of solutions – detailed information about the options considered and the reasons for making any recommendations.
(e) Planning and scheduling analysis – details of job breakdown, plans and schedules and estimates – including proposals from individual workers, allocation of tasks and scheduling of tasks, etc.
(f) Project costing – a full breakdown of the project costs and how the final price was calculated.
(g) Project justification – the reasons why the project will benefit the remote working organisation.
(h) Additional information – any supplementary information.

Appendix E: Risk assessment procedures

STEP 1

In which area of the business/project is there a threat?

STEP 2

For that area, identify each type of risk:

accidents .... Y/N
financial risks .... Y/N
labour risks .... Y/N
liability risks .... Y/N
marketing risks .... Y/N
natural disasters .... Y/N
political risks .... Y/N
sabotage .... Y/N
social risks .... Y/N
technical risks .... Y/N
other .... Y/N

STEP 3

For each 'Y'

— describe how the risk is likely to increase the chance of failure
— determine the best, worst and most likely outcomes
— for each outcome estimate the probability of its occurrence

on a scale of 1 to 10 (where 1 is 'highly unlikely' and 10 is 'highly probable')

— for each outcome assess the effect on the business or project on a scale of 'a' to 'j' (where a is 'mildly inconvenient' and j is 'certain failure of the project/business').

STEP 4

Consider the worst outcomes associated with each risk, then arrange the risks in order of priority (e.g. if eight risks have been identified, then the one with top priority will be '1/8'). Highest priority goes to those with a '10 j' rating, and lowest priority to those with a '1a' rating.

STEP 5

Starting with the highest priority, decide how to manage the risk in accordance with company strategy. Record your decision, and act accordingly (see Appendix F for sample Risk Management Record).

Appendix F: Risk Management Record

<table>
<tr><td colspan="2">Prepared by:</td><td>Date:</td></tr>
<tr><td colspan="3">Area of the business/project under threat:</td></tr>
<tr><td colspan="3">Type and description of risk:</td></tr>
<tr><td colspan="3">Best outcome:</td></tr>
<tr><td colspan="2">Best outcome probability:</td><td>Best outcome effect:</td></tr>
<tr><td colspan="3">Worst outcome:</td></tr>
<tr><td colspan="2">Worst outcome probability:</td><td>Worst outcome effect:</td></tr>
<tr><td colspan="3">Most likely outcome:</td></tr>
<tr><td colspan="2">Most likely outcome probability:</td><td>Most likely outcome effect:</td></tr>
<tr><td colspan="3">Risk Priority and Recommended Action:</td></tr>
<tr><td>Date:</td><td>By:</td><td>Action Taken:</td></tr>
</table>

Appendix G: Individual Progress Report – Guidelines

PURPOSE OF THE INDIVIDUAL PROGRESS REPORT

This report provides a mechanism for recording the time spent on an activity by a member of a project team, as well as comparing the actual time spent with the scheduled time. The benefits are twofold: first, that the progress of the project can be carefully monitored to enable deviations from original estimates to be detected as early as possible; and secondly, that the variations from initial elements can be evaluated and estimating procedures reviewed, in order that lessons learnt can be applied to future projects.

COPIES

This report is for internal use only. It is of the utmost importance that all individuals (including the project manager) complete a progress report on a predetermined and regular basis for every product on which they are working. Individuals should retain a copy for future reference, and a further copy should be made available to the project manager.

SECTIONS INCLUDED IN THE COMPLETED REPORT

(a) Name – of the team member.
(b) Project identifier.

(c) Week ending – date of the last day of relevant week.
(d) Product file name – to identify scheduled, deliverable component of the development to which the activities that follow relate.
(e) Activity code – to identify scheduled activity relating to the product under development.
(f) Days carried forward (to nearest half-day) – from previous weeks, for the activity specified.
(g) Days this week (to the nearest half-day) – spent on the activity specified.
(h) Days to finish (to the nearest half-day) – an estimate of the time required to complete the activity.
(i) Planned finish date – date by which the activity should be completed, as specified in the project schedule.
(j) Status – whether the activity will be completed early (E), on time (O), or late (L).
(k) Comments – any remarks relevant to completion of the activity.
(l) Additional information – this section should be used to record any details which may affect the success of the project, as follows: Activity code – to identify a scheduled activity to which the additional information relates; Comments, queries and problems – any relevant information, such as areas of uncertainty, problems and useful solutions, issues brought to light, reasons for delays, etc.

INDIVIDUAL PROGRESS REPORT

NAME:
PROJECT IDENTIFIER:
WEEK ENDING (SUNDAY):
PRODUCT FILE NAME:

Activity code	Days c/fwd	Days this week	Total to date	Days to finish	Planned finish date	Status E, O or L	Comments

PLEASE RECORD ADDITIONAL INFORMATION HERE:

Activity code	Comments, queries or problems

Appendix H: Project review – Summary

Project identifier:

STAGE	SUMMARY OF COSTS				SUMMARY OF EFFORT			
	Plan	Actual	Differs	Tot Dif	Plan	Actual	Differs	Tot Dif
1a. Conception								
1b. Feasibility								
2a. Requirement specification								
2b. Solution evaluation								
2c. Plan & Sched								
2d. Pricing								
2e. Proposal preparation								
3a. Proposal agreement								
3b. Proposal acceptance								
4. Design								
5. Creation								
6. Completion								
7. Success evaluation								
8. Performance appraisal								
9. Follow-on								

Appendix I: Configuration reporting – Guidelines

PURPOSE OF THE CONFIGURATION LIBRARY

The library maintains an up-to-date list of latest versions of all products released for a project.

COPIES

This report is for internal use only. It should be available to all members of a project team at all times during the development. A team member should update the log whenever they release a new version of a product (in addition to updating the Configuration Report for that product).

DATA TO BE INCLUDED ON THE COMPLETED REPORT

(a) Project identifier.
(b) Item name – to identify a product which has been released.
(c) Version number – two-digit serial number, identifying the latest version of the product.
(d) Released date – the day on which the version was released to other members of the team.
(e) Author – surname and initial of team member who produced this version.

Configuration Library
Project identifier:

Item – file name	Version no.	Released date	Author

PURPOSE OF THE CONFIGURATION REPORT

This report provides a mechanism for maintaining up-to-date information about all versions of a product.

COPIES

The report is for internal use only. Copies should be updated by the author of a product, when a new version is released and made available to all members of the project development team.

SECTIONS INCLUDED IN THE COMPLETED REPORT

(a) Project identifier.
(b) Product name – to identify the scheduled component to which the configuration report refers.
(c) Product description – narrative.
(d) Version number – two-digit serial number, identifying a version of the product.
(e) Author – surname and initial of team member who produced that version.
(f) Released date – date on which the version was released to other members of the team.
(g) Change reference – refers to a particular change to original specification.
(h) Location – any specific instructions for accessing this version.
(i) Associated products – reference to any other products which will be affected by, or are dependent on, this version.
(j) Comments – any remarks to clarify the content of this version.

Configuration Report
Project identifier:
Product file name:
Product description:

Version no.	Author	Released date	Change ref.	Location	Associated products
Comments:					
Comments:					
Comments:					
Comments:					
Comments:					
Comments:					
Comments:					

Appendix J: Task Sheet

PROJECT:			**TASK SHEET**
Task/Job Code:	Raised By:		Date:
Description:			
Quantity:			
Quality:			
Documentation:			
Dependency:			
Measurement/Completion Criteria:			
Resources/Support:			
Skills:			
Estimates/Name:	All:	Effort:	Duration/Availability:

Appendix K: Job breakdown – Summary

Task Identifier	Description	Dependency	Components	Milestone	Duration	Allocated
J22	Build workshop	J19 J20	T221 T222 T223 T224 T225		15 days	Various
T221	Lay foundations				3 days	J. Jones
T222	Build shell	T221	T2221 T2222	Completion of shell	5 days	Various
T2221	Build walls				3 days	J. Jones
T2222	Build roof	T2221			2 days	S. Smith
T223	Lay floor	T222			1 day	S. Smith
T224	Install services	T223			3 days	P. Price
T225	Workshop completion	T224	T2251 T2252 T2253	Completion of workshop	3 days	S. Smith

Appendix L: Summary of project plans

PRE-PROJECT ACTIVITY

1. Project Initiation

1a. Project conception

A project is conceived when a decision is made to resolve a problem, or take advantage of an opportunity.

Name	Effort (days)	Elapsed time	Cost	Comments

1b. Feasibility study

Perform an initial investigation of the business requirements and evaluate possible solutions.

Name	Effort (days)	Elapsed time	Cost	Comments

2. Preparation of Project Proposal

2a. Requirements specification

Identification of functional requirements – define, in detail, the business objectives.

Identification of technical requirements – define all technical aspects of the project.

Name	Effort (days)	Elapsed time	Cost	Comments

2b. Solution evaluation

Review of options – there are often several alternatives for satisfying the requirements of the project. What are they?

Solution assessment – evaluate the alternatives and make recommendations.

Name	Effort (days)	Elapsed time	Cost	Comments

Cost for this page	
Pre-project Total C/F	

2c. **Project planning and scheduling**

Job identification – what needs to be done and how? Identify alternatives; evaluate and select a preferred approach.

Role allocation – who is available/able to carry out each job?

Job estimation – what resources are required to complete each job to the required level of quality?

Job scheduling – when, and in what sequence, can jobs be carried out? Establish dependencies, identify key stages, milestones and deadlines. Use planning tools – e.g. Gantt charts, logic diagrams.

Name	Effort (days)	Elapsed time	Cost	Comments

2d. **Project pricing**

Project costing – what is the total cost, to the remote working organisation, of completing the project?

Evaluation of market options – who are the potential customers?

Pricing for customer – how much should each customer be charged for the product/service?

Name	Effort (days)	Elapsed time	Cost	Comments

2e. **Preparation of project proposal**

Project specification for client – produce a detailed description of the project objectives and how they can be achieved. Identify the benefits.

Internal proposal specification – includes more detailed breakdown of work, technical issues, etc.

Name	Effort (days)	Elapsed time	Cost	Comments

Cost for this page	
Pre-project Total C/F	

3. **Project Acceptance**

3a. **Proposal agreement**

Discuss and refine proposal

Name	Effort (days)	Elapsed time	Cost	Comments

3b. **Proposal acceptance**

Negotiate and approve contracts

Name	Effort (days)	Elapsed time	Cost	Comments

Cost for this page	
Pre-project Total C/F	

PRE-PROJECT PHASES – OTHER ACTIVITIES/COSTS

Specify the following:

Expenses

Equipment and Materials (for supplying to customer)

Supplies and Maintenance (for project development)

Equipment Hire (for project development)

Training of Team Members

Administration

Contingency

Labour:

Name	Effort (days)	Elapsed time	Cost	Comments

Other resources:

Resource	Source	Purchase/loan	Cost	Comments

PRE-PROJECT PHASES
SUMMARY

No. of man-days			
@ Rate per day			
Total			

Total Labour Costs	
Total Other Resource Costs	

TOTAL COSTS	
ELAPSED TIME	

PROJECT IMPLEMENTATION PHASES

4. Design of Products/services

Name	Effort (days)	Elapsed time	Cost	Comments

5. Creation of Products/services

Name	Effort (days)	Elapsed time	Cost	Comments

6. Project Completion

Name	Effort (days)	Elapsed time	Cost	Comments

Cost for this page	
Implementation Total C/F	

IMPLEMENTATION PHASES – OTHER ACTIVITIES/COSTS

Specify the following:
Expenses
Equipment and Materials (for supplying to customer)
Supplies and Maintenance (for project development)
Equipment Hire (for project development)
Training of Team Members
Administration
Contingency

Labour:

Name	Effort (days)	Elapsed time	Cost	Comments

Other resources:

Resource	Source	Purchase/loan	Cost	Comments

PRE-PROJECT PHASES
SUMMARY

No. of man-days			
@ Rate per day			
Total			

Total Labour Costs	
Total Other Resource Costs	

TOTAL COSTS	
ELAPSED TIME	

POST-PROJECT APPRAISAL PHASES

During the run-down period following completion of a project, it is essential to evaluate the success of the project and review the processes:

7. **Project Success Evaluation**

Evaluate project – did the project meet the expectations of customers, the remote working organisation and remote workers on time, within budget and to a satisfactory level of quality?

Identify learning points – identify strengths and weaknesses, improve organisation standards and procedures and identify training requirements.

Name	Effort (days)	Elapsed time	Cost	Comments

8. **Performance Review**

Appraise performance – of individuals and team.

Recognise commitment and performance – provide feedback/reward.

Name	Effort (days)	Elapsed time	Cost	Comments

9. **Follow-on Activities**

Identify opportunities – establish requirement for follow-on projects and identify marketing opportunities.

Maintain relationship with customer – recognise value of establishing a customer base, maintain contacts/customer information.

Name	Effort (days)	Elapsed time	Cost	Comments

Cost for this page	
Post-project Total C/F	

POST-PROJECT APPRAISAL PHASES – OTHER ACTIVITIES/COSTS

Specify the following:
Expenses
Equipment and Materials (for supplying to customer)
Supplies and Maintenance (for project development)
Equipment Hire (for project development)
Training of Team Members
Administration
Contingency

Labour:

Name	Effort (days)	Elapsed time	Cost	Comments

Other resources:

Resource	Source	Purchase/loan	Cost	Comments

POST-PROJECT
APPRAISAL – SUMMARY

No. of man-days			
@ Rate per day			
Total			

Total Labour Costs	
Total Other Resource Costs	

TOTAL COSTS	
ELAPSED TIME	

PROJECT SUM OF
PHASES

Total Labour Costs	
Total Other Resource Costs	

TOTAL COSTS	
ELAPSED TIME	

ADDITIONAL PROJECT ACTIVITIES/COSTS

Specify the following:
Overheads
Expenses
Equipment and Materials (for supplying to customer)
Supplies and Maintenance (for project development)
Equipment Hire (for project development)
Training of Team Members
Administration
Project Management
Insurance
Contingency

Labour:

Name	Effort (days)	Elapsed time	Cost	Comments

Other resources:

Resource	Source	Purchase/loan	Cost	Comments

FINAL PROJECT TOTAL

Total Labour Costs	
Total Other Resource Costs	

TOTAL COSTS	
ELAPSED TIME	

NOTE: This estimate does not include:

-
-
-

Appendix M: Milestone plans

Project identifier

No.	Milestone	Scheduled	Actual	Comments
M1				
M2				
M3				
M4				
M5				
M6				
M7				
M8				
M9				
M10				
M11				
M12				
M13				
M14				
M15				
M16				
M17				
M18				
M19				
M20				
M21				
M22				
M23				
M24				
M25				
M26				
M27				
M28				
M29				
M30				
M31				
M32				
M33				
M34				
M35				

INDIVIDUAL MILESTONE PLAN

Project identifier

Team member:

No.	Milestone	Scheduled	Actual	Comments

Appendix N: Evaluating computer software

The following checklist will be useful in assessing the value of computer software.

- What are the functional requirements of the software? Consider

  - what must it do (i.e. distinguish between essential and desirable facilities)?
  - why should it be done (i.e. the latest facilities may appear impressive but are they really of benefit to you)?
  - how should it be done (i.e. is there any requirement for specific tools, techniques or processes to be used, and if so, why and are there alternatives)?
  - who will use the facilities and when (i.e. will anyone need to be trained; how will it affect workloads)?
  - what volumes are anticipated (i.e. consider future expansion requirements)?

- What are the technical requirements of the software? Consider

  - must it run on existing hardware?
  - what maintenance/support will be required?

- Evaluate each option available. Consider

  - does it satisfy the essential functional requirements?
  - does it satisfy other requirements?

- how easy is it to use (i.e. what is the quality of the instructions/manuals)?
- how flexible is it?
- what is the quality of the output?
- are there any restrictions on volume/speed, etc.)?
- how reliable is it (i.e. what evidence is there)?
- will it run on existing hardware?
- will any enhancements to existing hardware/software be required?
- are any licences required?
- what support/maintenance is available?
- what are the costs (i.e. calculate all initial costs, such as installation, as well as running costs)?
- what is the availability of all components?
- what are the benefits?
- what are the drawbacks?

- Review all the options, including the possibility of improving manual systems. Make a decision and justify it!

Recommended reading

The following are examples of the many project management books available:

Brown, M. (1992), *Successful Project Management in a Week*, London: British Institute of Management.

Burton, C. (1994), *A Practical Guide to Project Management*, London: Kogan Page.

Lock, D. (1992), *Project Planner*, Aldershot: Gower.

Lock, D. (1996), *Project Management*, 6th ed., Aldershot: Gower.

Spinner, M. (1992), *Elements of Project Management*, New York: Prentice-Hall.

8 Legal Considerations

- Topics outline
- Employed or self-employed?
  Appendix A: Summary of differences
  Appendix B: Self-employed Status Checklist
  Appendix C: Specimen Agreement
  Appendix D: Contract for Services – completion explanation
- Provision of equipment
- Insurance
  Appendix E: Insurance Checklist
- Health and safety
- Data Protection Act 1984
  Appendix F: Registration Checklist
  Appendix G: Exemptions from registration
  Appendix H: Data protection principles
  Appendix I: Individual rights
  Appendix J: Enforcement

TOPICS OUTLINE

The topics outlined in this section are:

- Who should provide the equipment and how?
- What insurance will be required?
- What health and safety regulations will apply?
- Will we need to register under the Data Protection Act 1984?

EMPLOYED OR SELF-EMPLOYED?

1. **Does it matter whether workers are employed or self-employed?**
   Yes, it does, both for yourself and the worker.
   In practice, there are very significant differences between employed and self-employed status. See Appendix A to this chapter for Summary of the Differences.
2. **Does it matter how self-employed status is set up?**
   Yes, it does, particularly to yourself.
   If a person has been paid gross on a self-employed basis and it is later proved that he/she was in law employed, then the employer can be required to pay to the Inland Revenue income tax on the payments made and to the DSS both the employer's and the employee's NI contributions.
   In short, a mistake can be very expensive.
3. **Will a written agreement of self-employed status be sufficient?**
   No. A written agreement will certainly help; however, the description of a person in the agreement as self-employed

will not in itself be sufficient. The Inland Revenue may still argue that the person is really employed.

4. **What circumstances are important for self-employed status to be upheld?**

   The Inland Revenue will look at all the circumstances of the work to decide the issue. See Appendix B to this chapter for Checklist of the circumstances which point to self-employed rather than employed status.

5. **How can definite confirmation of self-employed status be obtained?**

   A ruling can be obtained from the Inland Revenue as to whether a proposed arrangement will constitute self-employed status.

   The ruling is given at local level and so is relatively easy to obtain.

   The DSS has indicated that they will accept an Inland Revenue ruling properly obtained.

   If you want to be safe rather than sorry, then obtaining a ruling from the Inland Revenue is a must.

6. **What form should a written agreement involving self-employed status take?**

   See Appendix C to this chapter for Specimen Agreement; and see Appendix D for Completion Explanation relating to the specimen agreement.

Appendix A: Summary of differences

	EMPLOYED	SELF-EMPLOYED
Income Tax	**Deductions under PAYE** The employer has to deduct income tax and pay it to the Inland Revenue under the PAYE Scheme.	**No PAYE** The employer pays gross. The burden of PAYE administration is avoided.
	Net payment The employee receives only the net payment.	**Gross payment** The employee receives the gross payment and does not have to pay any tax on it until after the end of the tax year in which it was earned.
	Restricted tax-deductible expenses The expenses which the employee can claim against his/ her income are self-restricted.	**Wider tax-deductable expenses** The rules for deduction of expenses incurred by a self-employed person are much more generous.
National Insurance (NI)	**Employer's contribution** The employer has to pay the relevant NI contribution for an employee.	**No employer's contribution** The employer has no NI contribution to pay for a self-employed person.
	Employee's contribution The employer also has to deduct the employee's NI contribution at the current rate.	**Self-employed contribution** The self-employed person pays a weekly flat rate and then an additional percent age onthe profits. The total may well be less than the employee's contribution and most of it is paid later.

Employment	**Employment rights**	**No employment rights**
Protection	An employee may have the benefit of all the statutory employment rights, including redundancy and unfair dismissal.	The statutory employment rights do not apply to a self-employed person.
Statutory Sick pay (SSP)	**Payment of SSP** The employer has to pay SSP to all qualifying employees. The payments cannot be reclaimed unless the employer is a 'small employer'.	**No payment of SSP** The employer has no requirement to make any payments to a self-employed person who is sick.

Appendix B: Self-employed Status Checklist

1. Does the worker use his/her own premises to carry out work?

2. Does the worker provide or pay for his/her own equipment and materials?

3. Does the worker control his/her own hours of work?

4. Does the worker have to a degree (if not complete), control of how the end-product of the work is achieved?

5. Is the worker free to accept and undertake work for other people, even competitors?

6. Does the worker have no guarantee of regular work being provided?

7. Does the worker receive no payment for holiday or sickness?

8. Does the worker bear whatever risks may be involved in the business?

If the answer is yes to most of the above questions, then the worker will indeed usually be self-employed.

Appendix C: Specimen Agreement

CONTRACT FOR SERVICES

Use the Completion Explanation in Appendix D to review the Agreement, and find out what each clause means; and decide whether you want to amend any of the clauses or add extra clauses.

PARTICULARS

'the Date' :

'the Business' :

'Address' :

'the Subcontractor :

'Address' :

'the Commencement Date' :

'the Termination Date'

(1) The date on which payment of the Fees is made following completion of the Services, *or*

(2) if earlier, *or*

(3) the date of termination under clause 12.

'the Services'	:
'the Fees'	:
'the Retention'	:
'the Release Date'	:

NB: In the following Agreement the above terms shall have the meanings as set out in these Particulars.

THE AGREEMENT

This Agreement is made on the Date

between

(1) the Business
(2) the Subcontractor.

1. **Definitions**

   In this Agreement the following expressions have the following meanings:

1.1	'the Quality Standards'	The current written standards of the Business designed to ensure proper methods of work and high quality of finished product.
1.2	'the Equipment'	Whatever computers, screens, keyboards,

		modems, computer accessories, stationery of any description, fax machines, telephone and other office equipment are required to provide the Services.
1.3	'the Payment Arrangements'	After submission of an invoice by the Subcontractor following the Termination Date, *and* after payment to the business by the relevant customer.
1.4	'the Rights'	The whole of the copyright and other rights in the nature of copyright in the material and products resulting from the performance of the Services by the Subcontractor under this Agreement.

2. **Interpretation**

2.1 **Headings**

The headings to clauses are inserted for convenience only and shall not affect the construction of this Agreement.

2.2 **Status**

Nothing contained in this Agreement shall be construed or have effect as constituting any relationship of employer and employee between the parties.

2.3 **Gender and singular/plural**

Where the context so requires or permits:

2.3.1 words expressed in any gender shall include any other gender;

2.3.2 the singular shall include the plural.

2.4 **Joint and several**

Where any party is more than one person:

2.4.1 that party's obligations in this Agreement shall take effect as joint and several obligations;

2.4.2 anything in this Agreement which applies to that party shall apply to all those persons collectively and each of them separately; *and*

2.4.3 the benefits contained in this Agreement in favour of that party shall take effect as conferred in favour of all those persons collectively and each of them separately.

3. **Commencement**

This Agreement starts on the Commencement Date and remains in force until the Termination Date.

4. **Duties**

The Business engages the Subcontractor to provide and the Subcontractor shall provide the Services for and on behalf of the Business.

5. **Good Faith**

The Subcontractor shall devote such part of his time and attention as is necessary to perform his duties under this Agreement properly and fully.

6. **Observe the Quality Standards**

The Subcontractor acknowledges and agrees that:

6.1 he has read the Quality Standards.

6.2 he will observe and perform the Quality Standards in providing the Services.

7. **Equipment**
   7.1 The Business may provide to the Subcontractor on its usual terms all or part of the Equipment.
   7.2 The Subcontractor shall provide for himself the Equipment to the extent that it is not provided by the Business.
8. **Insurance**
   The Subcontractor shall ensure at his own expense that he has adequate public liability insurance if required to do so by the Business.
9. **Fees and Payment**
   9.1 **Payment of the Fees**
   The Business shall pay to the Subcontractor the Fees as stated in the Payment Arrangements.
   9.2 **Retention**
   The Business may make the Retention from the Fees.
   9.3 **Release of the Retention**
   The Retention shall be released to the Subcontractor on the Release Date.
10. **Confidential Information**
   The Subcontractor shall both during this Agreement and after its termination keep confidential and not (except as authorised or required for the purposes of this Agreement) use or disclose or attempt to use or disclose to any person any confidential information relating to the organisation, finances, methods and business activities of and concerning the Business and its customers.
11. **Intellectual Property**
   11.1 **Belongs to the Business**
   The Rights shall belong to the Business absolutely.
   11.2 **Execution of documents, etc.**

At the request of the Business the Subcontractor will execute any assignments or other documents and will do any other thing required to give effect to clause 11.1.

12. **Termination on Default**

The Business may by written notice terminate this Agreement immediately and without liability for compensation or damages if:

12.1 the Subcontractor fails or ceases to perform his duties under this Agreement to the reasonable satisfaction of the Business.

12.2 any circumstances arise which give reasonable grounds in the opinion of the Business to believe that the Subcontractor has or may become unable or unsuitable to perform his duties under this Agreement.

12.3 the Subcontractor is guilty of any act which brings the Business into disrepute or which in the reasonable opinion of the Business is prejudicial to its interests.

12.4 the Subcontractor (if a company) convenes a meeting of its creditors or suffers a petition to be presented at a meeting to be convened or other action to be taken with a view to its liquidation except for the purposes of and followed by amalgamation or reconstruction.

12.5 a receiver or administrative receiver is appointed of any of the Subcontractor's property.

12.6 the Subcontractor dies or becomes bankrupt.

13. **Consequences of Termination**

13.1 **Return of possessions**

Within seven days of the Termination Date the

Subcontractor shall return to the Business any Equipment, documents and materials which are in his possession belonging or relating to the Business.

13.2 **Restriction**

For one year from the Termination Date the Subcontractor shall not provide services comparable to the Services for and on behalf of any person, firm or company who is or was a customer of the Business during the year preceding the Termination Date.

13.3 **Continuation of provisions**

The expiration or earlier termination of this Agreement shall not affect such of its provisions as are expressed to operate or have effect afterwards or are necessary to the proper fulfilment of this Agreement nor any right or action already accrued to either party in respect of any breach of this agreement by the other party.

14. **Variation**

No variation or amendment of this Agreement or oral promise or commitment related to it shall be valid unless committed to writing and signed by or on behalf of both parties.

15. **Previous Documents**

This Agreement is in substitution for any previous agreements express or implied between the parties which shall be terminated by mutual consent from the Commencement Date.

16. **Non-assignable**

The rights and obligations in this Agreement are personal to the parties and cannot be assigned or transferred.

17. **Notices**

17.1 **Method of service**

Any notice given under this Agreement shall be in writing and may be served:

17.1.1 personally;

17.1.2 by registered or recorded delivery mail;

17.1.3 by telex or facsimile transmission.

17.2 **Address for service**

Each party's address for the service of notice shall be the address stated in this Agreement or such other address as is specified in writing to the other party.

17.3 **Time of service**

A notice shall be deemed to have been served:

17.3.1 if it was served in person, at the time of service;

17.3.2 if it was served by post, 48 hours after it was posted;

17.3.3 if it was served by telex or facsimile transmission at the time of transmission.

18. **Jurisdiction**

This Agreement shall be governed by English law and the parties consent to the exclusive jurisdiction of the English Courts in all matters regarding it.

As witness the hands of the parties on the date first above written

Signed on behalf
of the *Business* }
in the presence of: ....

Signed by
the *Subcontractor* }
in the presence of: ....

Appendix D: Contract for Services – completion explanation

GENERAL

The document is designed such that all detail to be inserted is on the front page and the rest can be left as it is.

This, hopefully, avoids the tiresome need to go through *all* the document every time it is used.

PARTICULARS

The Date

Insert the date on which the Agreement is made.

The Business

This is you.

Insert the 'legal person' who is entering into the contract. If you are a limited company, this will be the company name.

If you are a partnership, this will be one or more of the partners' names or the partnership trading name.

If you are a sole trader, this will be your own name or your trading name.

You also insert your business address.

The Subcontractor

Again, insert the 'legal person' who is entering the Agreement. If the Subcontractor is a limited company, this will be the company name.

If the Subcontractor is a partnership, this will be one or more of the partner's names or the partnership trading name.

If the Subcontractor is a sole trader, this will be his/her name or the trading name.

NB: Should you ever need to sue on the Agreement, the 'person' against whom proceedings will be issued is the 'person' shown as the Subcontractor. This is one reason why it is important to get the information right from the start.

The Commencement Date

Insert the date from which the engagement of the Subcontractor is to start.

The Termination Date

If completion of the Services and payment of the Fees were the only Termination Date, then this could mean the Agreement going on and on.

To avoid this, a 'backstop' date needs to be included in (2). This date will be the last date by which you require the Subcontractor to complete the Services.

The Services

This is a very important part of the Particulars.

It is vital that the Services which you require from the Subcontractor are specified as clearly as possible. Problems can be avoided if the parties have a mutual understanding of what is to be achieved, and of the 'goalposts'.

The Services which you require may fall into categories, most elements of which remain the same on each occasion. If so, you should consider building up a set of precedents for describing 'the Services'. You could then either write this information from the relevant precedent into the Particulars (incorporating any details unique to the transaction), *or* you could use the precedent as a separate document annexed to the Agreement and enter here: 'As per the Schedule annexed'.

Similarly, if you use a separate 'Specification' to detail the services to be provided (perhaps because it is a lengthy document), you would enter here: 'As per the Specification dated . . . . . and number . . . . .'.

The Fees

This is an equally important part of the Particulars, since it sets out what you are to pay for the Services. Again, the aim is to make the payments required as clear as possible.

The Fees may be:

(a) one lump sum
(b) instalment payments
(c) a deposit and a final payment
(d) hourly or daily rates (with or without an estimate of the number of hours or days which it will take to complete the Services)
(e) some other method of calculation.

Remember to include any agreed provision about Expenses or other Extras.

Note

1. Clause 5 of the Terms makes it clear that the Fees are *exclusive* of VAT. Do *not* therefore include VAT in the figure or figures which you show as the Fees. However, you *can* show VAT as a separate figure. This will be a requirement if the Agreement is to act as the VAT invoice.

2. Clause 6 of the Terms provides for interest to be paid at the interest rate shown in clause 1.1.2 upon any late payments.

The Retention

Where you will be keeping a Retention from the Fees to secure against satisfactory completion of the Services, insert percentage or amount of the Retention (e.g. 5 per cent of the Fees).

Where you do not intend to make a Retention, delete 'the Retention' and 'the Release Date'.

The Release Date

Insert here the date on which the Retention will be released to the Subcontractor.

It might be an actual date, *or* it might be a 'formula' such as:
'six months from the Termination Date', *or*
'within 28 days of the date on which the Business receives payment in full from its customer for whom the Services by the Subcontractor are required'.

NB: the terms in the Particulars are used throughout the Agreement and will have the meanings which you have given to them. Therefore be *precise* and give full and sufficient meanings to the Particulars.

THE AGREEMENT

1. Definitions

EXPLANATION

It is good practice to set out at the beginning of a legal document the meanings to be given to particular words or phrases used in the document.

The clause can be regarded as 'the dictionary' of the contract.

Whenever these particular words are used in the Agreement, you need to look back to this section to find the meaning given.

ADDITION/AMENDMENT

None necessary, *unless* you add clauses to the Agreement and need to include here further definitions.

2. Interpretation

2.1 Headings

EXPLANATION

The headings are included for ease of reference. However, it is made clear that the words used do not alter the meaning of the actual clauses.

ADDITION/AMENDMENT

None necessary.

2.2 Status

EXPLANATION

It is here expressly stated that you are not the employer and the Subcontractor is not an employee. As mentioned earlier in the section, a statement of this kind is not sufficient in itself. You also need some or all of the other 'self-employed' features set out in the Self-employed Checklist in Appendix B, p. 270.

ADDITION/AMENDMENT

None necessary.

2.3 Gender and singular/plural

EXPLANATION

This clause means that one gender and the singular or

plural can be used throughout and yet be 'all-inclusive'. Otherwise cumbersome phrases like 'he', 'she', 'they' or 'it' or 'his', 'hers', 'theirs' or 'its' or 'person or persons' would be necessary.

ADDITION/AMENDMENT

None necessary.

2.4 **Joint and several**

EXPLANATION

This clause will only apply when the Subcontractor is a partnership.

The rights and obligations of the Agreement are enjoyed/owned collectively, *or* by each individual partner.

In other words, you can, for example, enforce the obligations against any one partner if you so choose. You do not have to enforce against all the partners. The decision will depend upon the circumstances at the time.

ADDITION/AMENDMENT

None necessary.

3. **Commencement**

EXPLANATION

This clause is, hopefully, self-explanatory.

See the Particulars for the meanings of 'Commencement Date' and 'Termination Date'.

ADDITION/AMENDMENT

If you want to change the Termination Date, e.g. the

Agreement is only to terminate when the Retention is released, then make the necessary amendment to the definition of 'the Termination Date' in the Particulars, e.g. to achieve the example given above:

(i) amend (1) to read:
'1. the Release Date'

(ii) leave (2) and (3) as they stand.

4. **Duties**

EXPLANATION
The clause sets out the Subcontractor's primary duty, i.e. to provide the Services.

ADDITION/AMENDMENT
None necessary

5. **Good Faith**

EXPLANATION
The Subcontractor can take on other work to do at the same time as your work, *but* this clause means that you can insist upon the Subcontractor spending sufficient time on your work to get it finished as agreed and in a proper fashion.

ADDITION/AMENDMENT
None necessary.

6. **Observe the Quality Standards**

EXPLANATION
This clause assumes that you have a separate written

'Quality Standards' specification or booklet. It ties in the subcontractor to reading and sticking to the Quality Standards.

ADDITION/AMENDMENT

If you do not have any written 'Quality Standards', then delete this clause and clause 1.1 in the Definitions, and substitute:

> '6. **Quality Standards**
>
> 'The Subcontractor shall observe and perform in relation to the Services such instructions and directions as the Business may give for the purpose of ensuring proper methods of work and a high-quality finished product.'

7. **Equipment**

EXPLANATION

If you choose, you can provide all or part of the required equipment for the Subcontractor. You are under no obligation to do so.

See the section titled 'Use of Equipment' for information about the terms upon which you should, or should not, provide equipment. Whatever equipment you do not provide, the Subcontractor has to provide.

ADDITION/AMENDMENT

None necessary.

8. **Insurance**

EXPLANATION

The clause mentions only public liability insurance. You can require the Subcontractor to have this.

See the precedent Agreements in the section titled 'Use of Equipment' for provisions about insurance of any equipment which you provide to the Subcontractor.

ADDITION/AMENDMENT
If you want to have the option to require the Subcontractor to take out indemnity insurance (i.e. to cover any liability for negligent work), then add the words 'and/or indemnity insurance' immediately after 'public liability insurance'.

9. Fees and Payment

9.1 Payment of the Fees

EXPLANATION
This clause sets out your primary obligation, i.e. to pay the Fees.

As already mentioned in relation to the Particulars, you must ensure that 'the Fees' are clearly and fully specified. It is the Particulars which contain the essential detail in this respect.

The 'Payment Arrangements' are defined in clause 1.3. The definition means that payment is made after you have received an invoice from the Subcontractor, and also after you have been paid by the customer.

ADDITION/AMENDMENT
If you want different arrangements for payment, then alter the definition in clause 1.3, e.g. 'within 28 days of the submission of an invoice by the Subcontractor', or 'within 28 days of approval of the work produced by the Services being given by the customer'.

9.2 Retention

EXPLANATION

The word 'may' used here means that you have a choice as to whether you will, or will not, make the Retention. You are not under an obligation to do so.

ADDITION/AMENDMENT

None necessary.

9.3 Release of the Retention

EXPLANATION

The clause is self-explanatory.

ADDITION/AMENDMENT

None necessary.

10. Confidential Information

This clause imposes an obligation of 'confidentiality' on the Subcontractor. It is obviously designed and intended to protect confidential information connected with your business which comes into the possession of the Subcontractor.

ADDITION/AMENDMENT

None necessary.

11. Intellectual Property

11.1 Belongs to the Business

EXPLANATION

See clause 1.4 for the definition of 'the Rights'. *Only* copyright has been included. This covers all original written work and original computer software.

The Rights will belong to you, so that you will be in legal possession to exploit them in the future.

ADDITION/AMENDMENT
If you want to include intellectual property rights beyond copyright, then delete the definition of 'the Rights' in clause 1.4 and substitute: 'patents, registered and unregistered designs, copyright (including without limitation computer programs) and all other intellectual property protection wherever in the world enforceable'.

11.2 **Execution of documents, etc.**

EXPLANATION
It is possible that you may need the Subcontractor to do something to enable you to protect the intellectual property rights (e.g. sign an application form). This clause imposes an obligation on the Subcontractor to do whatever you may require.

ADDITION/AMENDMENT
None necessary.

12. **Termination on Default**

EXPLANATION
There are a number of circumstances in which you will want to be able to terminate the Agreement immediately.

This clause sets out the 'Default events'. They essentially cover situations of breach of the Agreement, poor performance, financial difficulty and death.

ADDITION/AMENDMENT
None necessary.

13. Consequences of Termination

13.1 Return of possessions

EXPLANATION
The clause is self-explanatory.

ADDITION/AMENDMENT
None necessary.

13.2 Restriction

EXPLANATION
The clause is designed to protect against the Subcontractor 'stealing' your customers.

The Subcontractor must not provide similar services to any defined customer.

A defined customer is any customer of the Business during the one-year period immediately prior to the Termination Date.

ADDITION/AMENDMENT
None necessary.

13.3 Continuation of provisions

EXPLANATION
The clause does two things:

(a) It continues beyond the Termination Date the

operation of certain provisions. These provisions are those which are either said to so continue or whose continuation is necessary for the fulfilment of the Agreement.

(b) It makes clear that any accrued rights of action of either party can still be pursued after the Termination Date.

ADDITION/AMENDMENT

None necessary.

14. Variation

EXPLANATION

It is legally important that any variation of this Agreement or any promise relating to it should be clear. Often, if verbal variation or promise is relied upon, this is not so and dispute can ensue.

The clause therefore requires any such variation or promise to be in writing, *and* signed by the parties. Hopefully, this arrangement will minimise disputes.

ADDITION/AMENDMENT

None necessary.

15. Previous Documents

EXPLANATION

The clause makes clear that this Agreement supersedes any previous agreement or arrangements between the parties.

ADDITION/AMENDMENT

If there is another contract or Agreement with the Subcontractor which you want to continue, then:

(i) delete this clause entirely
(ii) renumber the following clauses.

16. **Non-assignable**

EXPLANATION

This clause means that neither party can pass on its rights and liabilities to another party.

ADDITION/AMENDMENT

If you do want a controlled right of assignment or transfer, then delete the present clause and substitute:

Assignment with Consent

'The rights and obligations in this agreement are personal to the parties and can only be assigned or transferred by one party with the written consent of the other.'

17. **Notices**

17.1 **Method of service**

EXPLANATION

Various methods of service of a notice are possible. This clause defines those which may be used for the purposes of this Agreement.

ADDITION/AMENDMENT

None necessary.

17.2 **Address for service**

EXPLANATION
The clause is self-explanatory.

ADDITION/AMENDMENT
None necessary.

17.3 **Time of service**

EXPLANATION
Legally, it can be important to establish when a notice has been served.

This clause states the 'deemed' time of service for the various methods of service.

ADDITION/AMENDMENT
None necessary.

18. **Jurisdiction**

EXPLANATION
The clause is self-explanatory

It prevents proceedings being taken in any other country.

SIGNATURES

The Business

EXPLANATION
The signature clause is worded for signature on behalf of a company or a partnership.

Where the Business is a company or a partnership, a director should sign his/her name and then write underneath 'Director'.

The signature of the director or partner is to be placed immediately to the right of the wording. The signature must be witnessed. The witness should sign immediately underneath the wording and then add his/her address and occupation, e.g.:

J.P. Bloggs
1 Nowhere Street
Somewhere
Secretary

ADDITION/AMENDMENT
If the Business is not a company but a sole partner, then:

(i) delete the present wording and substitute:

'*Signed* by
'the *Business*
' in the presence of ....   '

(ii) follow the directions for signature given in 'The Subcontractor'.

For explanation see below.

The Subcontractor

EXPLANATION
The signature clause is worded for signature by a sole trader.

He/she should sign immediately to the right of the wording.

The signature must be witnessed. The witness should sign immediately under the wording and then add his/her address and occupation (as illustrated, above).

ADDITION/AMENDMENT
If the Business is a company or partnership, then:

(i) delete the present wording and substitute:

'*Signed* on behalf of
'The *Subcontractor*
' in the presence of ....'

(ii) follow the instructions for signature given in 'The Business' explanation.

OTHER POSSIBLE CLAUSES

Warranty of authority

'Each of the parties warrants its power to enter into the Agreement and that it has obtained any necessary approvals to do so'.

EXPLANATION

It is important as far as the validity of any agreement is concerned to ensure that the parties have the legal capacity to enter into it.

The inclusion of the above clause would mainly be to 'concentrate minds' upon the need for capacity.

'Necessary approvals' would include, for example, resolutions of the company.

ADDITION/AMENDMENT

None necessary.

Power of Attorney

'The Subcontractor irrevocably appoints the Business or a designated representative or director of the Business to be his attorney to execute any document required to be executed by him under or in connection with this Agreement and the performance of it'.

EXPLANATION

You may need to have a document signed by the Subcontractor in connection with this Agreement, e.g. protect intellectual property rights under clause 11, *but* the Subcontractor may be unable or unwilling for whatever reason to sign. What do you do?

If you include this clause, you would be placed in the position of being able to sign the required document as power of attorney for the Subcontractor.

It is a handy provision to avoid the difficulty of a stalemate.

ADDITION/AMENDMENT

None necessary.

Arbitration

'All disputes or differences which shall at any time arise between the parties touching or concerning this Agreement or its construction or effect or the rights, duties or liabilities of the parties under or by virtue of it or otherwise or any other matter in any way connected with or arising out of the subject-matter of this Agreement shall be referred to a single arbitrator to be agreed upon by the parties or in default of agreement to be nominated by the President for the time being of the Chartered Institute of Arbitrators in accordance with the Arbitration Act 1950 or any statutory modification or re-enactment of it for the time being in force.'

EXPLANATION

Disputes between the parties which cannot be settled by agreement will have to be settled either by the courts or by arbitration.

The advantages of arbitration are:

(a) ability to choose the person and procedure to be used in resolving a dispute

(b) avoidance of publicity.

The disadvantages can be:

(a) cost (the arbitrator has to be paid)
(b) sometimes delay
(c) lack of interlocutory powers (e.g. grant of an injunction or order for disclosure of documents)
(d) exclusion of the courts (a court will not normally deal with a dispute where the agreement includes an arbitration clause)
(e) very limited rights of appeal against the arbitrator's decision.

This clause should be included where you prefer dispute resolution by arbitration rather than the courts.

ADDITION/AMENDMENT
None necessary.

NB: Remember if you do add any clauses, you *must* check and alter the clause numbering as appropriate throughout the Agreement.

PROVISION OF EQUIPMENT

WHO SHOULD PROVIDE THE EQUIPMENT AND HOW?

1. Alternatives.

2. Supply own equipment.

3. Loan the equipment free of charge

- 3.1 Consumer credit and sale of goods not applicable
- 3.2 Display screen regulations applicable
- 3.3 Specimen Loan Agreement.

4. Lease/rent the equipment
   - 4.1 Consumer Credit Act 1974
     - 4.1.1 The 1974 Act not applicable
     - 4.1.2 Specimen Rent/Hire Agreement
     - 4.1.3 The 1974 Act applicable.
   - 4.2 Supply of Goods and Services Act 1982
     - 4.2.1 In the course of a business
     - 4.2.2 Merchantable quality
     - 4.2.3 Minimising liability for merchantable quality
     - 4.2.4 Fitness for purpose.
   - 4.3 Health and Safety at Work Act 1974
     - 4.3.1 Obligations as supplier
     - 4.3.2 Obligations as a person 'conducting a business'.
   - 4.4 Other possible liabilities
     - 4.4.1 Dangerous goods
     - 4.4.2 Unsafe goods.

5. Sell the equipment
   - 5.1 Consumer Credit Act 1974 not applicable
   - 5.2 Sale of Goods Act 1979
   - 5.3 Health and Safety at Work Act 1974
   - 5.4 Other possible liabilities.

1. Alternatives

The contractor/worker will need the required equipment to carry out the work.

There are several alternative ways for the equipment to be supplied:

(a) the contractor/worker can supply and use his/her own equipment.
(b) you can loan the equipment to the contractor/worker free of charge.
(c) you can lease/rent the equipment to the contractor/worker.
(d) you can sell the equipment to the contractor/worker.

2. **Supply own equipment**
The advantage is that you will be free from all legal requirements relating to the equipment, whether health and safety, consumer credit, sale of goods, etc. But you will need to have some means of checking that the equipment of the contractor/worker will be adequate for the work involved.

3. **Loan the equipment free of charge**

3.1 **Consumer credit and sale of goods not applicable**
Because no payment is made, you will not be subject to the consumer credit or sale of goods legislation.

3.2 **Display screen regulations applicable**
However, if the equipment consists of or includes a workstation, you will have to comply with the Health and Safety (Display Screen Equipment) Regulation 1992.

See section 8, clause 4.4, for a brief summary of the requirements, and/or obtain from HMSO the following pamphlet: 'Display screen equipment – guidance on Regulations', L26 (ISBN 0 11886331 2).

3.3 **Specimen Loan Agreement**

It is prudent to have a written agreement setting out the terms of the loan, even though no payment is involved.

4. **Lease/rent the equipment**

4.1 **Consumer Credit Act 1974**

4.1.1 **The 1974 Act not applicable**

The 1974 Act will apply to a lease/rent agreement, unless:

(a) the one who takes the equipment on hire is a company; *or*

(b) the total amount involved is over £15 000; *or*

(c) the agreement is for less than three months.

4.1.2 **Specimen Rent/Hire Agreement**

Do not use the Specimen Rent/Hire Agreement in Appendix G if the 1974 Act applies.

4.1.3 **The 1974 Act applicable**

The requirements of the 1974 Act are complicated and very important. Failure to 'get it right' may mean that the agreement will be unenforceable.

You should either avoid the application of the 1974 Act, or if this is not possible or desirable, seek professional advice upon the form of agreement of use and the procedural requirements which must be observed.

4.2 Supply of Goods and Services Act 1982

Where goods are hired out in the course of a business, there are implied terms of merchantable quality or fitness for purpose.

4.2.1 In the course of a business

Your business is to provide programming or other computer work. You then hire out a workstation. Is this 'in the course of a business'?

If the hiring is a 'one-off', then there is a strong argument that it is not in the course of a business (i.e. of hiring workstations), so that the 1982 Act will not apply.

However, if hiring occurs on more than one occasion, it is more likely that such hiring would be held by a Court to be in the course of a business, so that the 1982 Act will apply.

4.2.2 Merchantable quality

Where the 1982 Act applies, goods hired must be of merchantable quality.

The legal meaning of the expression 'merchantable quality' is not straightforward. There have been a number of cases involving debate over the meaning.

In simple terms, the goods must:

(a) be fit for their usual or ordinary purpose; *and*
(b) work properly; *and*
(c) be in a satisfactory condition.

4.2.3 Minimising liability for merchantable quality

You can minimise the potential liability for merchantable quality if, before the hiring:

(a) you point out to the hirer any defect in the goods (you will not then be liable for those defects); *and*

(b) you insist that the hirer carefully examine the goods (you will not then be liable for the defects which such examination should reveal).

4.2.4 Fitness for purpose

The purpose for which goods are required may be:

(a) *implied* – i.e. the purpose for which goods of that kind are usually bought; *or*

(b) *express* – i.e. the hirer makes known a particular purpose.

The goods must be fit for that purpose, implied or express.

4.3 Health and Safety at Work Act 1974

4.3.1 Obligations as supplier

Where you hire goods, you will be 'the supplier' of the goods.

The 1974 Act imposes obligations on the supplier of articles (which includes equipment or appliances) for use at work.

The obligations may be summarised as follows:

(a) to make sure the article is safe and without risk to health so far as reasonably practicable;
(b) to carry out any necessary examination;
(c) to supply as far as reasonably practicable adequate information about the correct use of the article.

4.3.2. **Obligations as a person 'conducting a business'**

The 1974 Act requires an employer to conduct his/her business in such a way as to ensure, so far as reasonably practicable, that persons not in their employ who may be affected by the employer's activities are not exposed to risks to their health or safety.

The hirer of goods will be 'a person not in your employ' who will be affected by your business activities.

Where the goods are or include a workstation, this means that the Health and Safety (Display Screen Equipment) Regulations 1992 will apply.

See section 8, clause 4.4, for a brief summary of the requirements of the Regulations, and/or obtain from HMSO the following pamphlet: 'Display screen equipment – guidance on Regulations', L26 (ISBN 0 11 886331 2).

4.4 Other possible liabilities

4.4.1 Dangerous goods

In certain circumstances, you could be liable for damage wholly or partly caused by a defect in the product/goods which you have supplied. Part I of the Consumer Protection Act 1987 contains the relevant legislation.

4.4.2 Unsafe goods

If the product/goods supplied are unsafe, you will be guilty of an offence. 'Unsafe' means when there is a risk, above a minimum risk, of death or personal injury being caused by:

(a) the goods;
(b) keeping, using or consuming the goods;
(c) assembling the goods;
(d) any emission or leakage from the goods;
(e) reliance on the accuracy of any measurement, calculation or other reading made by or by means of the goods.

Part II of the Consumer Protection Act 1987 contains the relevant legislation.

5. Sell the equipment

5.1 Consumer Credit Act 1974 not applicable

Provided that the price is payable in not more than three instalments over a period of less than 12 months, the Consumer Credit Act 1974 will not apply.

This means that you will *not* need to comply with all the requirements as to form of agreement and procedure for completion.

5.2 **Sale of Goods Act 1979.**

Where goods are sold in the course of a business, there are implied terms of merchantable quality or fitness for purpose.

See 4.2.1 in this section for the meaning of 'in the course of a business'.

See 4.2.2 in this section for the meaning of 'merchantable quality'.

See 4.2.3 in this section for the meaning of 'minimising liability for merchantable quality'.

See 4.2.4 in this section for the meaning of 'fitness for purpose'.

5.3 **Health and Safety at Work Act 1974**

Where you sell goods, you will be the 'supplier' of the goods. Consequently, the same obligations are placed upon you as were mentioned where you hire out the goods/equipment.

See 4.3.1 in this section for 'Obligations as supplier'.

See 4.3.2 in this section for 'Obligations as a person conducting a business'.

These obligations will again be applicable.

5.4 **Other possible liabilities**

The position is the same as where you hire out the goods/equipment.

See 4.4.1 in this section for possible liability for 'Dangerous goods'.

See 4.4.2 in this section for possible liability for 'Unsafe goods'.

INSURANCE

What insurance will be required?

1. **Basic purpose**

   The basic purpose of insurance is to compensate when you suffer loss.

   The types of risk which can be insured are very varied.

2. **Checklist**

   The Checklist in Appendix E to this sections shows:

   (a) the types of insurance available
   (b) the risks which they cover
   (c) whether they are likely to be relevant to you or not.

3. **Combined insurance**

   It is possible to combine a number of risks in a 'package' or 'combined' insurance policy.

4. **Advice**

   This section mentions only very briefly those types of risk for which it is considered you might need insurance.

   However, the best step you can take, in practice, is to discuss your requirements with a good insurance broker.

Appendix E: Insurance Checklist

Liability Insurance (provides indemnity against a legal liability)

1.1 Employer's liability	Risk of a claim by an employee for bodily injury or disease (*Compulsory* where you have employees)	Possibly applicable
1.2 Public liability	Risk of a claim by a member of the public for injury or damage sustained	Probably applicable
1.3 Professional liability	Risk of a claim by a client/customer for alleged loss or injury caused by a negligent service	Possibly applicable
1.4 Product liability	Risk of a claim for damage or injury caused by a product	Probably *not* applicable
1.5 Directors and officers	Risk of a claim for liability breaches of duty of trust	Possibly applicable
1.6 Goods in transit	Risk of a claim for loss or liability damage to goods being transported	*Not* applicable

1.7 Motor liability	Risk of a claim for damage, death or injury sustained in a car accident (third party)	Possibly (*Compulsory*)
	Also risk of damage cover is to your own vehicle (comprehensive)	(*Compulsory*)

Perils Insurance (provides cover against a particular peril)

2.1 Fire	Risk of damage to property or contents by fire, lightning, explosion and a number of other causes	Possibly applicable
2.2 Theft	Risk of damage and loss through theft and burglary	Possibly applicable
2.3 Plate glass	Risk of breakage to plate glass windows	Possibly applicable
2.4 Business interruption	Risk of loss of revenue through enforced cessation of business for a period	Possibly applicable
2.5 Money	Risk of loss of money when being held or in transit	Probably *not* applicable
2.6 Travel	Risk of various losses and expenses	*Not* applicable

	expenses connected with travel (baggage, medical expenses delay, cancellation etc.)	(unless travelling abroad)
2.7 Fidelity	Risk of loss through the dishonesty of an employee	Probably *not* applicable

Personal Insurance (provides cover against personal risks)

3.1 Personal accident	Risk of temporary or permanent incapacity, serious injury or death from an accident	Optional *but* advisable
3.2 Permanent health	Risk of loss of or reduction in income because of an accident	Optional *but* advisable
3.3 Critical	Risk of a serious illness	Optional
3.4 Whole of life	Risk of death	Optional *but* advisable

Provision Insurance (provides funds upon certain events)

4.1 Medical expenses	Provision for medical expenses upon illness or hospitalisation	Optional
4.2 Pension	Provision for retirement	Optional *but* advisable

4.3 Endowment	Provision at the expiration of a stated period or upon earlier death	Optional *but* advisable
4.4 Term	Provision upon death occurring during a stated period	Optional

HEALTH AND SAFETY

What health and safety regulations will apply?

1. **The legislation**

   The law on health and safety is contained in the following:

   (a) Factories Act 1961
   (b) Offices, Shops and Railway Premises Act 1963
   (c) Health and Safety at Work Act 1974
   (d) Regulations made under the Acts
   (e) Health and Safety Executive Codes of Practice and Guidance Notes.

2. **Factories Act 1961**

   The Act applies only to factories.

   'Factory' is defined to cover premises in which persons are employed in manual labour for the purpose of gain. A list of types of manual labour included is given in the Act. In essence, the Act covers manufacturing premises.

   The Act is *not* of relevance to persons working at home.

3. **Offices, Shops and Railway Premises Act 1963**
The Act, as its title states, applies to office premises and shop premises.

It is expressly stated that the Act does not apply to:

(a) premises where only self-employed people work; *or*
(b) a dwelling.

Also the definition of 'an employee' is confined to a person employed under a contract of service or of apprenticeship. It does not extend to a self-employed person supplying personal services.

The Act is therefore *not* of relevance to persons working at home.

4. **Health and Safety at Work Act 1974**

4.1 **Scope of the Act**
This Act lays down broad obligations to be performed and observed by employers and others.

4.2 **Obligations to employees**
'An employee' is again confined to a person employed under a contract of service or of apprenticeship. It does not extend to any self-employed persons.

Therefore none of the obligations contained in the Act towards employees is relevant.

4.3 **Obligations to others**
Under the Act, there is an obligation for all employers to conduct their business in such a way as to ensure, so far as is reasonably practicable, that persons not in their employ who may be affected by the employer's activities, are not exposed to risks to their health or safety.

This obligation is wide enough to extend to customers, contractors, visitors, etc.

A particular example of this obligation will be where a workstation is supplied to the person for use in carrying out the required work at home. The Health and Safety (Display Screen Equipment) Regulations 1992 would apply.

4.4 Display screen regulations

The Health and Safety (Display Screen Equipment) Regulations 1992 came into operation on 1 January 1993.

There is a transitional period up to 1 January 1997 for existing workstations to meet the minimum health and safety requirements.

'Display screen equipment' includes any type of computer display screen.

'Workstation' means the whole computer assembly, including the work desk and chair, work surface and immediate work environment.

'Operator' refers to a self-employed person, as opposed to an employee (who is called a 'user'), who habitually uses a display screen for a significant part of their normal working day.

Where you supply the workstation, you are required by the regulations:

(a) to assess the risks to health and safety of an operator using the workstation by 'performing a suitable and sufficient analysis of the workstation'.

(b) where risks to health and safety are identified, to

reduce them to the lowest extent reasonably practicable.

(c) to ensure that workstations first put into service after 1 January 1993 comply with the minimum health and safety standards set out in the regulations.

(d) to provide adequate information to the operator about the health and safety aspects of the workstation and any measures taken as a result of the workstation assessment or of the minimum health and safety requirements.

For further information: Obtain the following leaflet from HMSO: 'Display Screen Equipment – Guidance on Regulations', L26 (ISBN 0 11 886331 2).

DATA PROTECTION ACT 1984

Will we need to register under the Data Protection Act 1984?

1. **What is the purpose of the Act?**
   The Act was passed to give protection and rights to individuals about whom information is recorded on computer.

   The Act places obligations on those who record and use personal data. They must register and follow sound and proper practices.

   The Act also gives the right to individuals to find out information recorded about them, to challenge it when appropriate and, in certain circumstances, to claim compensation.

2. **Do we need to register?**

See Appendix F to this chapter and work through the Registration Checklist to see if you need to register.

Then see Appendix G to make sure that you are not covered by any of the exemptions from registration.

In summary, if you hold personal information about living individuals on a computer, you will need to register.

3. **How do we register?**

An application for registration must be made on Form DPR1 or DPR4.

The forms, together with Notes and various other helpful literature, can be obtained from:

Office of the Data Protection Registrar
Wycliffe House
Water Lane
Wilmslow
Cheshire
SK9 5AF
Tel: (Enquiries) 01625 545745

Form DPR1 is in two parts: Part A and Part B. Only one copy of Part A needs to be completed. However, a separate Part B needs to be completed in respect of each purpose for which personal data is held (the Notes give a list of 70 possible purposes!).

But for most small businesses, Form DPR4 will be more appropriate. This one form covers the four main purposes for which registration is usually required by such businesses:

(a) Personnel/employee administration.
(b) Marketing and selling.
(c) Purchase/supplier administration.
(d) Customer/client administrator.

4. **How much does registration cost?**
The current registration fee for an application is £75.

5. **What rules does registration mean that we have to observe?**
The rules to be observed are called Data Protection Principles. There are eight rules: see Appendix H to this chapter for a summary of the eight Principles.

6. **What rights does an individual have under the Act?**
The individual's rights are:

(a) Access
(b) Compensation
(c) Correction or erasure
(d) Complaint.

See Appendix I to this chapter, if you require further information on or about Individual Rights.

7. **What happens if we do not comply with the Act?**
The Registrar is responsible for enforcement of the Act. In cases of non-compliance, he can choose between:

(a) An enforcement notice.
(b) A de-registration notice.
(c) A transfer prohibition notice.
(d) Prosecution in the criminal courts.

Appendix J to this chapter contains more about enforcement.

8. **Where can we obtain more information?**
The Office of the Data Protection Registrar (see above for address and telephone number) has produced eight Guidelines which seek to explain the Act as clearly as possible, and a Guidance Note series which deal with specific issues in more detail. You can obtain this information by contacting the Office.

(Reference to the Guidelines has been made in compiling this summary of the Data Protection Act.)

Appendix F: Registration Checklist

Do I need to be registered?

- **Data User**

Do you:

1. have a computer
2. on which information is held
3. about individuals
4. who are alive
5. and who are identifiable?

If you have answered Yes to all these questions, then you are a data user and need to be registered.

- **Computer Bureau**

Do you:

1. have a computer
2. which you use for others or allow others to use
3. for processing data
4. whether or not for payment?

If you have answered Yes to all these questions, then you are a computer bureau and need to registered.

Appendix G: Exemptions from registration

1. Personal data held by an individual for domestic or recreational purposes. (The exemption does *not* extend to a company or organisation.)

2. Information that the law requires to be made public.

3. Where exemption is required to safeguard national security.

4. Personal data held only for payroll, pensions and accounts purposes.

5. Personal data held by an unincorporated members club which relate only to members of the club.

6. Personal data which consist only of names and addresses to be used for mailing purposes.

Note: The above list represents a simplification of the exemptions (their precise requirements and extent are rather more complicated). If you think that any of the exemptions might apply, seek legal advice to confirm.

Appendix H: Data protection principles

1. 'The information to be contained in personal data shall be obtained, and personal data shall be processed, fairly and lawfully.'

Examples of information unfairly obtained are where deception or unfair pressure has been used.

2. 'Personal data shall be held only for one or more specified and lawful purposes.'

'Specified purposes' means those which have been registered. You should therefore make sure to register all your required purposes and update by further registration as and when necessary.

3. 'Personal data held for any purpose or purposes shall not be used or disclosed in any manner incompatible with that purpose or those purposes.'

You can comply with this Principle by ensuring that your register entries fully describe all uses and disclosures of personal data (the forms include checklists to assist you).

4. 'Personal data held for any purpose or purposes shall be adequate, relevant and not excessive in relation to that purpose or those purposes.'

You should only collect the right amount and type of information required about individuals.

5. 'Personal data shall be accurate and where necessary, kept up to date.'

Data will need updating where its contents may affect a decision; for example, whether or not to grant credit or confer a benefit.

6. 'Personal data held for any purpose or purposes shall not be kept for longer than is necessary for that purpose or those purposes.'

You will need to review your personal data regularly and delete information which is no longer required.

7. 'An individual shall be entitled:

   (a) at reasonable intervals and without undue delay or expense

      (i) to be informed by any data user whether he holds personal data of which that individual is the subject; *and*
      (ii) to have access to any such data held by a data user, and

   (b) where appropriate, to have such data corrected or erased.'

For more details on individual rights see Appendix I to this chapter.

8. 'Appropriate security measures shall be taken against unauthorised access to, or alternation, disclosure or destruction of, personal data and against accidental loss or destruction of personal data.'

You are required by this Principle to have a security policy in respect of personal data.

Appendix I: Individual rights

1. **Access**

   An individual can request a data user to say whether that data user holds personal data about the individual and to supply a copy.

   A request should be in writing.

   The data user can charge a fee for dealing with a request.

   The maximum fee allowed is £10.

   The data user must respond within 40 days of receiving the request.

2. **Compensation**

   An individual can claim compensation from a data user if he/she suffers damage as a result of inaccurate personal data held, or of a breach of security about him/her by the data user.

   An application for compensation has to be made to the Court, not the Registrar.

   'Damage' includes financial loss or physical injury but not distress. No compensation can be claimed for distress alone.

   If the data user can prove that all reasonable care was taken to ensure the accuracy of the personal data, or to prevent the destruction, disclosure or access in question, no compensation is payable.

3. **Correction**

   If personal data is inaccurate (i.e. incorrect or misleading about any matter of fact), the individual may apply to the

Court for an order that the data user should correct or erase the data.

4. **Complaint**

An individual who considers that there has been a breach of the Data Protection Principles or of any provision of the Act may complain to the Registrar.

The Registrar has to notify the individual of the result of his investigation and of any action which he proposes to take.

Appendix J: Enforcement

1. **An Enforcement Notice**
   This Notice states:

   (a) the Principle or Principles concerned.
   (b) the breaches alleged.
   (c) the steps to be taken by the data user to comply.
   (d) the time within which the steps are to be taken.
   (e) the right of appeal to the Data Protection Tribunal.

   Failure to comply with an enforcement notice is a criminal offence.

2. **A De-registration Notice**
   This Notice states:

   (a) the Principle or Principles involved.
   (b) the breaches alleged.
   (c) why an enforcement notice would not be adequate.
   (d) what details the Registrar proposes to remove from the Register.
   (e) the period after which the removal will be completed.
   (f) the right of appeal to the Data Protection Tribunal.

3. **A Transfer Prohibition Notice**
   This notice may be served where it appears to the Registrar that a registered data user proposes to transfer personal data to a place outside the United Kingdom.

   Contravention of such notice is a criminal offence.

4. **Criminal prosecution**

Non-registration and certain breaches of or failures under the Act can be prosecuted as criminal offences.

The maximum fine is £2000.

Index